CONTENTS

FOREWORD

Will I ever do anything without the pressure of a deadline?

Paradoxically this preface is the last thing to be written yet one of the first to be read.

I'm trying to write it on a swaying tube train. Late again. A few people scattered through the carriage, rocked on their seats, gazing desultorily at one another, and at the same advertisements. What brings us here for the first and only time in our lives, synchronised in this twenty-minute fragment of the earth's existence? On one advertisement a high-speed train drives through the shattered faces of three station clocks: 'Red Star gets you there like there's no tomorrow'. Will there be a tomorrow? This kind of reverie is common enough – maybe we were all thinking it simultaneously on that train: at once the banality of time and its infinite mystery.

This book and the six films from which it has grown have been made by many people. Some of us had previously worked with each other and came together deliberately to make the series. But I also have the feeling that in the long and complicated gestation of the films as much happened by accident as by design. Members of the original working group altered, new relationships were made, and old ones suspended; our assumptions were challenged by things haphazardly heard or seen and chance meetings gave rise to new ideas; schedules collapsed, plans went awry, but something better came along to redeem the chaos with order; carefully thought-out sequences fell out of the films in favour of half-considered images whose meaning was only found through the process of editing.

We did not start with a thesis which we wanted to prove, but with questions we wanted to ask. Often both the films and this book take the form of a dialogue.

The structure of both films and book are inevitably time-related. While this conditions what can be said and how, we have tried to make the forms of each responsive to the themes, using images and words not just to illustrate and explain ideas, but to reach for them poetically. Metaphor is important because through it both meaning and mystery are brought together in their most potent form.

It is often said that film is an inappropriate medium in which to play with ideas. I take an opposite view: perhaps 'playing with ideas' is just what film is quite good at doing. In our series it allows the intellectual world to sit alongside the world of industrial work; it lets us contrast the dreams of advertising with lived experience, menstrual dreams with dreams of death, and poetry with everyday conversation; images of quantum physics, astrology, ancient Holy Days, modern holidays, love life, work life, poignant May Day songs from Padstow, eerie fire festivals from Lewes ... all can coexist on a television screen in the corner of every room.

In a film, words and images are laid end to end on a ribbon of celluloid. As the film is shown it unfolds in sequential time. But in spite of the linear nature of this process, the hope of our series is that it makes new interconnections, across the time of the film, which challenge the prevailing view that time itself is linear.

Some people who have been near to death have said that at that moment the whole of their lives passed before them in a flash. I often dream of a film in which for one crazy moment every image, thought, word, feeling and note of music will be present at one and the same moment. This feeling I have about film relates to something John Berger has written:

'As for myself, ever since I was born on November 5th, 1926, I have struggled with my puzzling conviction that everything is simultaneous.'

This thought, and the thoughts of many other contributors, are present in this book, which was made by Chris Rawlence (with whom I directed the films) and Irène Rado-Vajda. But it is impossible to disentangle the growth of these ideas from the contributions of so many people in the making of the films and the book: Sophie Balhetchet, Ray Beckett, Alistair Cameron, Ray Cornwall, Mick Coulter, Christopher Cox, Greg Dupré, Penny Forster, Rosalind Haber, Terry Hardy, Celia Lowenstein, Daniel McCormack, John Midgeley, Andy Nelson, Jo Nott, Di Ruston, Howard Sharp, Caroline Spry, Sarah Stinchcombe, Nancy VandenBergh, Dai Vaughan, Glenn Wilhide. And many others, not least those whom we love and live with.

Michael Dibb

ABOUT TIME

BASED ON
THE CHANNEL 4 TELEVISION SERIES
DIRECTED BY MICHAEL DIBB AND CHRISTOPHER RAWLENCE

EDITED BY
CHRISTOPHER RAWLENCE

DESIGNED BY
IRÈNE RADO-VAJDA

JONATHAN CAPE
THIRTY-TWO BEDFORD SQUARE LONDON
IN ASSOCIATION WITH CHANNEL 4 TELEVISION COMPANY LTD

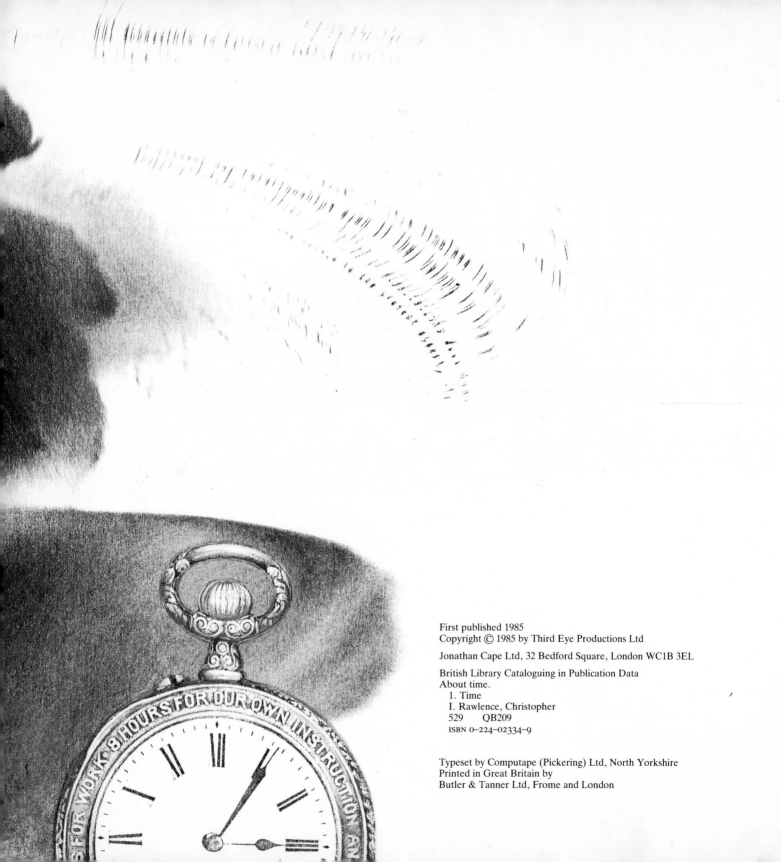

First published 1985
Copyright © 1985 by Third Eye Productions Ltd

Jonathan Cape Ltd, 32 Bedford Square, London WC1B 3EL

British Library Cataloguing in Publication Data
About time.
  1. Time
  I. Rawlence, Christopher
529   QB209
ISBN 0–224–02334–9

Typeset by Computape (Pickering) Ltd, North Yorkshire
Printed in Great Britain by
Butler & Tanner Ltd, Frome and London

PREFACE

Accurate clock time – the time by which we organise ourselves – has become so indispensable to industrial society, which is founded on the regular beat of its rhythm, that we've come to look on it as *the* time. The final pip of the time signal, automically monitored, seems to indicate the time of the universe itself ... precisely.

But human experience contradicts this powerful idea. The contrasting rhythms of heartbeat, digestion, the pattern of waking and sleeping, the reproductive cycle and our responses to the turning seasons are some of the cohabiting times of our biology, just as the times of dreams, memories, boredom, sex and absorption each have their own rhythms in consciousness. These human realities, which we share with each other through art, love, co-operation and just living together, suggest that there are many dimensions to time in the universe. Yet the authority of clock time in our culture, underwritten by science, is so absolute that the evidence of human experience is stigmatised as too subjective.

This book is about righting the balance. It is about how our obsession with one kind of time has made us prisoners of it – a confinement that involves the danger-ous repression of many of the essential rhythms that define humanity. Dangerous, because the physical and psychological consequences of undervaluing innate rhythms, or denying them sufficient expression, perhaps nourishes the dark side of human nature at the expense of the light.

Put another way, Western culture is characterised by a separation of mind and body that goes back at least as far as the Enlightenment and the thought of Descartes. Our hope here is to find ways of talking about time that bridge this gap, so reuniting the world of the intellect with the biological, emotional and spiritual reality of being human.

This is why nearly all the contributors to this book root their discussion of time in human reality, putting lived experience before abstract philosophical speculation. In my links and introductions I have gone even further in this direction, basing many of them on personal memories.

This book is complementary to the Channel 4/Third Eye television series *About Time*. With a single exception, each section corresponds to a film in the series. The first section is by John Berger. The other four are based on conversations that we filmed with contributors to the programmes.

Christopher Rawlence

ONCE UPON A TIME

Once there was a young man who said to himself, 'This story about everybody having to die, that's not for me. I'm going to find a place where nobody ever dies.' And so he said goodbye to his parents and his family, and he set off on a journey.

And after several months, he meets an old man, with a beard down to his chest, trundling rocks off a mountainside in a wheelbarrow. He says to the old man, 'Do you know that place where nobody ever dies?'

The old man says, 'Stay with me, and you won't die until I have carted all this mountain away in my wheelbarrow.'

'How long will that take?'

'Oh, at least a hundred years.'

'No,' says the young man, 'I'm going to find that place where nobody ever dies.'

He travels on, and he meets a second old man, with a beard down to his waist. This old man is on the edge of a forest which seems to go on for ever and ever. And he's cutting branches off a tree.

The young man says, 'I'm looking for that place where nobody ever dies.'

'Stay with me,' says the old man, 'and you won't die until I've cut off the branches of every tree in this forest.'

'How long will that take?'

'At least two hundred years.'

'No. I'm going to find that place where nobody ever dies.'

He travels on again. He meets a third old man, with a beard down to his knees, and this old man is watching a duck drinking sea water from an ocean.

'Do you know that place where nobody ever dies?'

And the old man answers, 'Stay with me and you won't die until this duck has drunk the whole ocean.'

'How long will that take?'

'Oh, at least three hundred years, and who wants to live longer than that?'

'No,' says the young man, 'I'm going to find that place where nobody ever dies.'

The young man goes on. And he comes to a castle. The door opens and there is an old man, with a beard reaching down to his toes.

'I'm looking for that place where nobody ever dies.'

'You've found it,' replies the old man.

'Can I come in?'

'Yes, I would be glad, very glad, of company . . .'

Time passes. And one day the young man says, 'You know, I'd like to go back – just for a moment. I won't stay long but I just want to go back to say hello to my parents and to see where I was born.'

The old man says, 'Centuries have gone by, they're all dead.'

'I'd still like to go back if only – if only to see the street where I was born.'

So the old man says, 'All right, follow my instructions carefully. Go to the stables, take my white horse, a horse who is as fast as the wind, and never get off him. If you get off that horse, you'll die.'

The young man mounts the horse and rides away. After a while he comes to the beach where the duck was drinking the sea. The sea-bed is now as dry as a prairie. The horse stops at a little heap of white bones – all that is left of the old man with the beard down to his knees.

'How right I was not to stop here,' says the young man to himself. And he goes on, and he comes to where he saw the forest. The forest is now pasture-land – not a single tree is left.

'How right I was not to stop here,' says the young man to himself. The young man goes on and comes to where the mountain had been, but it is now as flat as a plain. For the third time he says to himself: 'How right I was not to stop here!'

Finally he arrives at the town where he was born. He recognises – nothing. Everything has changed. He feels so lost that he decides to go back to the castle. One day on his return journey, towards nightfall, he sees a cart drawn by an ox. The cart is piled high with old, worn-out boots and shoes. As he passes, the carter cries out, 'Stop, stop! Please get down. Look, a wheel of my cart is stuck in the mud. I'm alone. Please help me.'

The young man answers back, 'I'm in a hurry, I won't stop and I can't get off my horse.'

The carter pleaded, 'It will be dark in a moment, it'll freeze tonight, I'm old and you're young, please help me.'

So, out of pity, the young man got off his horse. Before his second foot was out of the stirrup, the carter grasped him by the arm and said, 'Do you know who I am? I am Death. Look in the cart at all those boots and shoes I have worn out chasing you! Now I have found you. Nobody ever escapes me . . .'

In this story we know at the beginning what the end will be. We know that the young man, however cunning, will eventually have to die. We know better than he does, we see through his illusions. We follow the story, not because we believe that he may escape death, but because we want to learn more about his cunning and, in response, Death's cunning.

When this story is being told there are at least four different time-perspectives or tenses in play. There is the *present* of the narrative – the present which the young man is living. There is the historic *past* – all this happened a while ago. There is the infinite *future*, the promise of which prompts the young man to try to outwit death. And then there is the time of the listener's imagination which has already seized the *whole* and all its tenses. I mention this now only in passing; we will come back to it later.

The story, first told centuries ago near Verona in Italy, still speaks to us. The inexorability of time, the inevitability of death, the desire for immortality – none of this has changed. Yet something has changed. The original listeners to the story, recognising the vanity, the uselessness of the young man's efforts, would probably have thought of him as being obstinately short-sighted. They would have seen him as an opportunist. Today we would simply say that he lacked realism, that he was a kind of dreamer. The difference between calling him *short-sighted* or calling him *unrealistic* is the difference between two approaches towards the enigma of time.

The original listeners believed that beyond time there was the timeless. The hero of the story is *short-sighted* because he has forgotten this, and his forgetfulness makes him an opportunist. In our culture the timeless does not exist. There is no rational place for it; which is why we call the story's hero unrealistic and think of him as a dreamer-victim.

Here is another story. It happened, a few years ago, in the next village.

If Jean were an animal, what kind of animal would he be? I'd answer: an elephant. He's not so large, but he has an extraordinary memory, he's strong, he is good-natured, he's secretive and he is gentle. Some people make the mistake of taking advantage of him, but only once, for when Jean goes berserk it's not something people forget easily. He works in a sawmill, and he lived once with a woman, about ten years older than him, whom he called – both in public and in private – Mother.

Mother had run away from her husband and children in the northern coalfields, and had come to work as a servant in the local hotel. There, one morning, Jean noticed her because she was crying whilst picking beans in the vegetable garden behind the hotel kitchen. She said she couldn't bear the thought of returning to the north, her husband beat her, could she stay with Jean in the village? If he took her in, she would cook for him.

11

She wore short dresses and had pretty legs. She cried and laughed without premeditation. Although fifty years old, it was a little as if she had just left school. Jean told her she could stay with him. For four years they lived together as man and wife.

Then she fell ill. She lost her memory. She could no longer pick beans or untie a knot. If she went down to the village, she couldn't find her way home. Without any reason, she accused Jean of being unfaithful to her. She became incapable of dressing herself properly. Now it was Jean who cooked for her.

The doctors finally named her illness 'premature senility'. To Jean it seemed as if she was returning to childhood. Then they told him her blood pressure was too low and that she must be sent to hospital. He accompanied her. As soon as they arrived in the hospital ward she began to cry; she sensed that everything was slipping away from her.

'Just slip away,' the nurse said to Jean, 'don't let her know you are leaving.'

Two days later he went to visit her. She was tear-stained and cowed like a prisoner. The specialist in his white coat behind an oak desk told Jean that her illness was incurable. There was nothing to be done. She would get worse and worse and the illness would kill her.

Jean pondered the matter for a long while. He thought about it when he was sawing wood and when he was sleeping in the bed they used to share. One day he decided: 'I'm going to fetch Mother on Saturday afternoon.'

The young doctor on duty at the hospital refused to give the permission for Mother to leave.

'Who are you anyway?' he asked Jean.

'I'm the only person she has in the world.'

'You are not even a relative, and you're not her husband.'

'I'm taking her away.'

'It's I who give orders here!' the doctor shouted.

'No. About Mother you have no right. And you can listen to me. I'm taking her home.'

'Get out of my sight – both of you!'

This is how the doctor admitted his defeat. Mother was the first to understand what this admission meant, and her whole face became radiant. Jean took her back to the ward and dressed her. He put on her black stockings – although it was the month of June – because she said she wanted to look smart.

'Are you hungry?' he asked her in the car.

'It was like a prison inside there.'

'Don't think about it.'

'They locked me in my room.'

'Don't think about it.'

'They treated me so badly.'

'I'm taking you home,' said Jean, 'and I've cooked us tripe for lunch.'

'I haven't eaten for a week.'

'We'll eat some tripe and this afternoon we can sleep.'

At the door of their house, the dog jumped up to lick her face. Whilst Mother was in hospital Jean had taken his dog to the sawmill with him each day. 'Imagine being alone in the house all those hours!' he said. Mother put her arms round the dog's shoulders, and Jean arranged the armchair in front of the open window. It was sunny and the plum trees were still in bloom.

'Sit there, Mother,' he said, 'I have something to fetch.'

As soon as she was alone, she got out of the chair and peered into the bedroom – the house had only two rooms. On the table by the bed Jean had arranged a vase of white lilac. The bedspread was lace. Jean came in from the garden with four kittens, two in each hand.

'I kept them all for you,' he said, 'I didn't drown a single one this time.'

She sat down and, taking a kitten in each hand, rubbed each one against her cheek. The other two Jean placed on her lap in the hollow between her legs.

'Our worries are over,' he said and kissed her forehead.

'Four kittens!' she cried with delight.

'You're home,' he said.

'Jean! How happy we are!'

He picked up a cloth for opening the oven of the wood stove.

'Doesn't it smell good?' she said. 'I didn't eat for a week.'

'You're free now, Mother, and we're together again.'

'For always!'

'Always.'

Within a month or two she no longer even recognised him.

●

The two stories – the first from Verona and the second from the village next to mine – are different in kind. The first is a fable, the second is about an event I myself witnessed. Yet they have things in common. In the fable there is the idiosyncratic human detail – for example, the fact that the young man finally gets off his horse out of *pity* for the old man. And in the second story there is something a little fabulous – the fact that Jean with his low status nevertheless insists that the doctor agrees to Mother's discharge from the hospital.

Both stories are about the human desire to discover or invent strategies for outwitting time. The young man in the fable tries to cheat death, whilst Jean and Mother set out to live a Saturday as if it will never be superseded.

A need for what transcends time, or is mysteriously spared by time, is built into the very nature of the human mind and imagination. One has to live with this need without deceiving oneself.

Time is created by events. In an eventless universe there would be no time. Different events create different times. There is the galactic time of the stars, there is the geological time of mountains, there is the life time of a butterfly. There is no way of comparing these different times except by using a mathematical abstraction. It was man who invented this abstraction. He invented a regular 'outside' time into which everything more or less fitted. After that, he could, for example, organise a race between a tortoise and a hare and measure their performances using an abstract time unit (minutes).

'Time is not an empirical conception. For neither coexistence nor succession would be perceived by us, if the representation of time did not exist as a foundation a priori . . . Time is nothing else but the form of the internal sense, that is, of the intuitions of self and of our internal state.'

Kant: *Critique of Pure Reason*

The problem, however, lies elsewhere; it arises from the fact that man himself constitutes two events. There is the event of his biological organism, and in this he is like a tortoise or a hare. And there is the event of his consciousness and imagination. Each of these events creates its own time and the two times are different. There is the time of his body and the time of his mind. The first time 'understands' itself, which is why hares and tortoises have no philosophical problems. The second time, the time of consciousness, has been understood in different ways in different periods. It is the first task of any culture to propose an explanation of the time of consciousness: of the relation between past, present and future, realised as such.

The explanation offered by European thought in the nineteenth century proposes a single, abstract, uni-linear law of time which applies to everything that exists, including consciousness. This explanation, whose task is to 'explain' the time of consciousness, treats that consciousness as if it were as inanimate as an extinct sun. If modern European man has become a victim without hope of his own positivism, the story starts here.

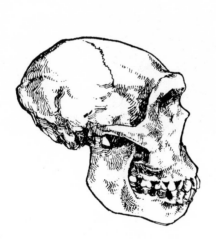 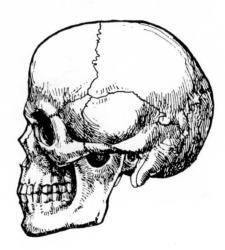

Man is always between two times. Hence the distinction made in all cultures – except the positivist one – between body and soul. The soul is first, and above all, the locus of another time, distinct from that of the body.

Everyone knows that stories are simplifications. To tell a story is to select. Only in this way can a story be given a form and so be preserved. The story-teller is like a dressmaker cutting a pattern out of cloth. She or he cuts from the cloth as fully and intelligently as possible. Inevitably there are narrow strips and awkward triangles which cannot be used; these have no place in the form of the story. If you tell a story about somebody you love, a curious thing happens. Suddenly you realise it is those strips, those useless remnants, which you love most. The heart has to retain all.

15

'I see the past, present and future', wrote Blake, 'existing all at once before me.'

The great contemporary Lebanese poet, Adonis, has written:

>Do you remember the house
>ours alone
>among the olive trees and figs
>with the source sleeping tight
>against the orphans of the eye
>
>Do you remember why the woods
>beat their wings like butterflies
>the earth's first night . . .
>
>The night . . .
>
>Drill wells in your breast
>be labyrinth and take me in

Rembrandt painted Hendrickje Stoffels, who was the great love of his life, many times.

One of his most modest paintings of her is, for me, one of the most mysterious. She is in bed. I think the picture was painted a little before the birth of Cornelia, Hendrickje's daughter with Rembrandt. Hendrickje and Rembrandt have lived together, as man and woman, for twenty years. In about two years' time Rembrandt will be declared bankrupt. Ten years earlier, Hendrickje came to work in his house as a nurse for his two-year-old son from his previous marriage. Hendrickje will die, although younger than Rembrandt, six years before him.

It is late at night; she has been waiting for him to come to bed. He has just entered the room. She lifts up the curtain of the bed with the back of her hand. The face of its palm is already making a gesture, preparatory to the act of touching his head or his shoulder, when he bends over her. In this portrait of Hendrickje she is entirely concentrated on Rembrandt's sudden appearance. In her eyes we can read her portrait of him. The curtain she is holding up divides two kinds of time: the daytime of the daily struggle for survival, and the night-time of their bed.

Rembrandt must have painted the picture partly from memory, recalling the times he had come to bed late and Hendrickje had been waiting for him. To a degree, then, it is a remembered image. Yet what it shows is a moment of anticipation – hers, and his by implication. They are about to leave the world of debts and monthly payments for the instantaneity of desire, or sleep or dreams. The picture was painted three hundred years ago. We are looking at it today. There is such a tangle of times within this image that it is impossible to *place* the moment it represents. It resists the mechanism of clocks or calendars. But this does not make it less real.

The image of Hendrickje in bed was preserved thanks to Rembrandt's art. A similar need to preserve images is felt by everyone at certain moments.

In this photograph by the great Hungarian photographer, André Kertesz, we see a mother with her child looking very intently at a soldier. We do not, I think, require the caption in order to know that the soldier is about to leave. The greatcoat hung over his shoulder, his hat, the rifle, are evidence enough. And the fact that the woman has just walked out of a kitchen and will shortly return to it is indicated by her lack of outdoor clothes. Some of the drama of the moment is already there in the difference between the clothes the two are wearing. His for travelling, for sleeping out, for fighting; hers for staying at home.

The caption reads: 'A Red Hussar leaving, June 1919, Budapest'. How much that caption means depends upon what one knows of Hungarian history. The Habsburg monarchy had fallen the previous summer. The winter had been one of extreme shortages (especially of fuel in Budapest) and economic disintegration. Two months before, in March, the Socialist Republic of Councils had been declared, under Béla Kun. The Western Allies in Paris, fearful lest the Russian, and now the Hungarian, example of revolution should spread throughout Eastern Europe and the Balkans, were planning to dismantle the new Republic.

19

A blockade was already imposed. General Foch himself was planning the military invasion being carried out by Rumanian and Czech troops. On June 13th, 1919, Clemenceau sent an ultimatum by telegram to Béla Kun, demanding a Hungarian military withdrawal, which would have left the Rumanians occupying the eastern third of their country. For another six weeks the Hungarian Red Army fought on but was finally overwhelmed. By August, Budapest was occupied. The first European Fascist regime, under Horthy, had been established.

If we are looking at an image from the past and we want to relate it to ourselves, we need to know something of the history of that past. And so the above information is relevant to the reading of Kertesz's photograph.

Everything in it is historical: the uniforms, the rifles, the street corner by the Budapest railway station, the identity and biographies of all the people who are (or were) recognisable – even the size of the trees on the other side of the fence. Yet it also concerns a resistance to history: an opposition.

This opposition is not the consequence of the photographer having said, 'Stop!' It is not because the resultant static image is like a fixed post in a flowing river. We know that in a moment the soldier will turn his back and leave; we presume that he is the father of the child in the woman's arms. The significance of the instant photographed is already claiming minutes, weeks, years. The opposition exists in the parting *look* between the man and the woman.

Their look, which crosses before our eyes, is holding in place what *is*, not specifically what is there around them outside the station, but what *is* their life, and what *are* their lives. The woman and the soldier are looking at each other so that the image of what *is* shall remain for them. In this look their being is opposed to their history, even if we assume that this history is one that they accept or have chosen.

How close a parting is to a meeting!

Everywhere in the world people have invented stories about the Beginning, about how the universe was created. All mythologies are a way of coming to terms with the fact that man lives between two times. He is born and he dies, like every other animal; yet he can imagine the origin and the end of everything. And as a result of this imagining, he lives with the eternal, with that which preceded time and will follow it, with that which is continually there behind time. The Red Hussar and his wife, like two people feeling their way round the furniture in a pitch-dark room, are looking for a door which opens on to what is behind time.

The Adams and Eves
continually expelled
and with what tenacity
returning at night!

Before,
when the two of them
did not count
and there were no months
no births and no music
their fingers were unnumbered.

Before,
when the two of them did not count
did they feel
a prickling behind the eyes
a thirst in the throat
for something other than
the perfume of infinite flowers
and the breath of immortal animals?
In their untrembling sleep
did the tips of their tongues
seek the bud of another taste
which was mortal and sweating?

Did they envy the longing
of those to come after the Fall?

Women and men still return
to live through the night
all that uncounted time.

And with the punctuality
of the first firing squad
the expulsion is at dawn.

Twenty years ago I saw a photo by Lucien Clergue, one of a series of photographs of a woman in the sea. For millenniums, the look of a woman's body in water has changed no more than the look of the waves. In the Egypt of the Pharaohs, in ancient China, this image could have been seen. Yet its unchanging quality represents only an instant when placed in the context of another time-scale.

Scientists now estimate that life first appeared on this planet four thousand million years ago. The story began with the coming into being of single organic molecules. These were produced, mysteriously, by the action of ultra-violet light on certain mineral elements. Then, after many many trials – and it's difficult to avoid the notion of a will – the first single organic cells produced something capable of reproduction.

Thus it can be said that it took four thousand million years to produce the form and surface of the human body. The camera recorded this image in less than one hundredth of a second. Do these incomprehensible figures touch the mystery of the sight of the figure in the sea? In a sense they do. What they cannot touch is the *urgency* of the image. What impresses or moves us always has an urgency about it.

In Clergue's photograph the sense of urgency can be explained by the nature of human sexuality and by cultural conditioning. It can be put into its place within the history of our biological and social evolution. Yet such an explanation remains unsatisfactory because, once again, it ignores the fact that consciousness cannot be reduced to the laws of uni-linear time. It is always, at any given moment, trying to come to terms with a whole.

From the idea of the whole come all the variants of the notion – ranging from astrology to the Koran – that everything that happens has already been conceived, has already been told.

There's another Italian story, this time from the Piedmont.

One day a farmer was on his way to the town of Biella. The weather was foul and the going was difficult. The farmer had important business, and so he went on, despite the driving rain.

He met an old man and the old man said to him, 'Good day to you. Where are you going in such a hurry?'

'To Biella,' answered the farmer, without slowing down.

'You might at least say, "God willing".'

The farmer stopped, looked the old man in the eye and snapped, 'God willing, I'm on my way to Biella. If God isn't willing, I've still got to go there all the same.'

The old man happened to be God. 'In that case,' said the old man, 'you'll go to Biella in seven years. In the meantime, you'll jump into that pond and stay there for seven years!'

Abruptly the farmer changed into a frog and jumped into the pond. Seven years went by. The farmer came out of the pond, smacked his hat back on his head, and continued on his way to Biella. After a short distance he met the old man again.

'Where are you going to in such a hurry?' asked the old man.

'To Biella,' replied the farmer.

'You might at least say, "God willing".'

'If God wills it, fine,' answered the farmer, 'if not, I know what is going to happen, and I'll jump into the pond of my own free-will!'

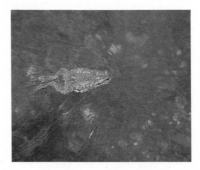

What I find wise in this story is the space it offers for the coexistence of free-will and a form of determinism. The 'God willing' offers a promise of a whole. The 'God willing' also, of course, poses insuperable problems. But to close this contradiction is more inhuman than to leave it open. Free-will and determinism are only mutually exclusive when the timeless has been banished, and Blake's 'Eternity in love with the productions of time' has been dismissed as nonsense.

No painter knew more about ageing than Rembrandt. It seems to me that throughout the course of a life the relation between space and time sometimes changes in a way that is revealing.

To be born is to be projected – or pulled – into time/space. Yet we experience duration before we experience extension. The infant lives minutes – without being able to measure them – before he lives metres. The infant counts the temporary absence of his mother by an unknowable time unit, not by an unknowable unit of distance. The cycles of time – hunger, feeding, waking, sleeping – become familiar to the infant long before any means of transport. Most children face the experience of being lost in time before the experience of being lost in space. The first vertigo – and the rage it produces – is temporal. Likewise time brings 'arrivals' and rewards to the infant before space does.

This is all the more striking if one compares a baby with a newborn animal. Many animals – especially those which graze – begin, within minutes of being born, a spatial quest which will continue all their lives. The comparative immaturity of the human baby, which obliges a long period of physical dependence, but which also permits a much longer and freer learning potential, encourages the primacy of time over space. Space is to animals what time is to sedentary man.

This human predisposition is linked with the early development of the human will. The will develops slowly – along with the ability to stand up, to walk, to talk, to name what one wants. But the will exists from birth onwards. (It is born from that expulsion.) In infancy the will selects very few objects from the full display of the existent. These few isolated objects, invested with desire or fear, then provoke *anticipation* and are thus delivered to the realm of time.

With age some people become more contemplative. Yet what is contemplation if it is not the acknowledgment, the celebration, of the spatial and the forgetting of the temporal? (For which the pre-condition is the abandonment of the personal will.) Within the act of contemplation the spatial does not exclude time, but it emphasises simultaneity instead of sequence, presence instead of cause and effect, ubiquity instead of identity.

25

Curiously, as I write these words I do not think of a passive hermit, but of a dead friend, who died because of his active politics.

His name was Orlando Letelier, a Chilean. He liked music. He was witty and debonair. And he had extraordinary courage. He had been a minister in Allende's government. When Allende was murdered and Pinochet seized power, Orlando and many others were arrested, held in prison, and tortured. Eventually Orlando was released on the condition that he left Chile. He travelled around the world, talking about the fate of his country. He encouraged other exiles. He persuaded certain governments to reduce or to stop their trade with Chile. He was eloquent and effective. Pinochet decided that Letelier had to be got rid of. His secret police organised the planting of a bomb in his car in Washington. One morning, when Orlando was driving to his office with a young American woman who was acting as his secretary, his car blew up. The woman was killed, and Orlando, after he had lost both legs, died.

The day after receiving the news of his assassination, I wrote the following:

Once I will visit you
he said
in your mountains
today
assassinated
blown to pieces
he has come to stay
he lived in many places
and he died everywhere
in this room
he has come between the pages
of open books
there's not a single apple
on the trees
loaded with fruit this year
which he has not counted
apples the colour of gifts
he faces death no more
there's not a precipice
over which his corpse
has not been hurled
the silence of his voice
tidy and sweet as the leaf of a beech
will be safe in the forest
I never heard him speak
in his mother tongue
except when he named the names
of patriots
the clouds race over the grass
faster than sheep
never lost
he consulted the compass of his heart
always accurate

took bearings from the needle of Chile
and the eye of Santiago
through which he has now passed.

Before the fortress of injustice
he brought many together
with the delicacy of reason
and spoke there
of what must be done
amongst the rocks
not by giants
but by women and men
they blew him to pieces
because he was too coherent
they made the bomb
because he was too fastidious
what his assassins whisper to themselves
his voice could never have said
afraid of his belief
in history
they chose the day
of his murder.

He has come
as the season turns
at the moment of the blood-red rowan-berry
he endured the time without seasons
which belongs to the torturers
he will be here too
in the spring
every spring
until the seasons returning
explode
in Santiago.

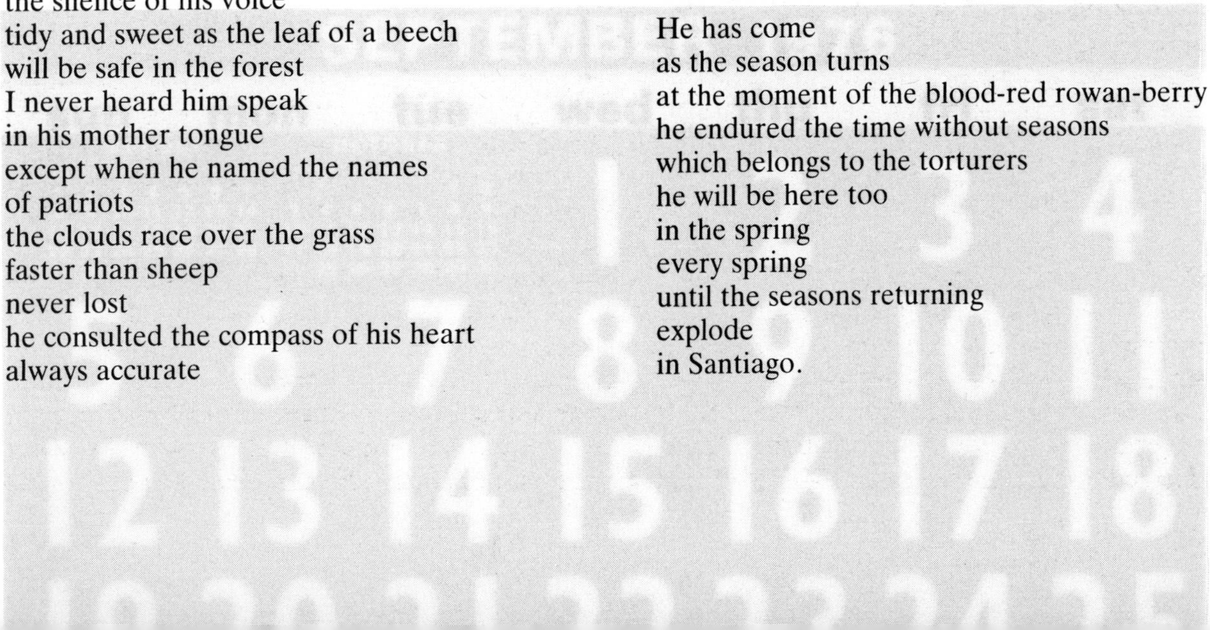

27

To the questions I ask about time, I have no answers. I only know that such questions are of the human essence and that any system of habits, or of reasoning, which dismisses them does violence to our nature. Since I first started writing I have been labelled a Marxist – a convenient category for those who so label me, and sometimes a shelter for myself. I am convinced by Marx's understanding of the role and mechanisms of capitalism. Within the world historical arena, the fighting is mostly as he foresaw. The questions to which I give voice here come from outside that arena.

I began with a story; I will end with story-telling.

Imagine a character in a story trying to conceive of his origin, trying to see beyond what he knows of his destiny, before the end of the story. His inquiries would lead him to hypotheses – infinity, chance, indeterminacy, free-will, fate, curved space and time – hypotheses not dissimilar to those at which we arrive when interrogating the universe.

The notion that life – as lived – is a story being told, is recurrent, expressed in multiple forms of religion, popular proverbs, poetry, myth, and philosophical speculation. Nineteenth-century pragmatism rejected this notion and proposed that the laws of nature were ineluctably mechanical. Recent scientific research tends to support the assumption that the universe and its working processes resemble those of a brain rather than a machine.

What separates the story-teller from his protagonists is not knowledge, either objective or subjective, but *their* experience of time in the story he is telling. (If he is telling his own story the same thing separates him as story-teller from himself as the subject.) This separation allows the story-teller to hold the whole together; but it also means that he is obliged to follow his protagonists, follow them, powerlessly, *through* and *across* the time which they are living and he is not. The time, and therefore the story, belongs to them. Its meaning belongs to the story-teller. Yet the only way he can reveal this meaning is by telling the story to others.

A story is seen by its listener or reader through a lens. This lens is the secret of narration. In every story the lens is ground anew, ground between the temporal and the timeless.

●

TIME IS MONEY

My first real job was as a builder's labourer in the graveyard of a small medieval church where I had to dig several soakaways for the drainage of rain water from the church roof. These holes in the ground had to be six feet deep and five feet across. The pay was £10 a week, which worked out at about two shillings (10p) an hour.

The first day was appalling. It was the first time that I had ever been faced with relentlessly boring and very hard work, and it felt as if it was going to go on for ever. All morning I went at the hard ground with my spade, making little impression. I looked up at the clock at what I thought were about ten-minute intervals, to realise that only two minutes had passed. By noon I had badly blistered hands and the dinner break felt like a sacred refuge. By the end of the afternoon I was still only one foot down into my first soakaway, locked into a hopeless encounter with unyielding yew roots. The prospect of the next day made me feel grey and nauseous.

On the third day the going got suddenly softer. By late afternoon I was through the roots and four feet down. Simply through shifting more soil, the day seemed to have passed more quickly.

On the fourth day the work got interesting. About midday and five feet down, my spade scraped something white. Bending down into the hole, I picked gently with my fingers at the earth that clung to the whiteness. Slowly the unmistakable curve of a human cranium emerged. I was standing in a grave.

Throughout the afternoon I unearthed more and more bones: a shoulder blade, some shattered ribs, scattered vertebrae, broken arms, and a whole thigh bone. Then the second skull appeared, and the thigh bones of five legs. By the end of the next day I had assembled the bones of six skeletons, all of which had been intermingled in the same small space.

The absence of rotting timber and corroded coffin handles suggested to me that these people had been put here a long time ago. The texture of the bones seemed to confirm this. They felt very old, and the worms had long since filled the empty spaces left by eyes and brains with compacted soil. It occurred to me that I was disinterring a medieval plague grave. What else would explain so many bodies, seemingly thrown in on top of each other?

My sudden involvement with the job was sanctioned by the vicar who asked that as far as possible the bones be kept intact. He wanted them respected so that he could re-bury them when the soakaways were completed. This meant I could take my time. The art was to get a whole skull out without breaking it, which was difficult because the bone was brittle. The vicar also wanted to re-bury the skeletons individually, which involved the challenge of matching the right bones to the right body. From the mindless digging of a hole, my work had been transformed into a skill. For a day and a half I was so completely absorbed at the bottom of the damp soakaway that I forgot about time and the dragging of the church clock hands.

This part of the book is about work, time, money and the roots of our society's obsession with clocks. Our conversation with Mike Cooley runs down consecutive right-hand pages. My piece about clock time runs in parallel down the left-hand pages.

MIKE COOLEY holds qualifications in engineering and a Ph.D. in computer-aided design. Author of many works on technological change, he holds a number of visiting professorships in Britain and Europe. For many years a senior design engineer in the aerospace industry, he holds several patents for systems he designed. An active trade unionist, he was National President of the Designers' Union AUEW-TASS in 1971. In 1981 he was joint winner of the $50,000 Alternative Nobel Prize for his work on socially useful products and human centred systems. He is married with two children.

LONDON COUNTY COUNCIL

14 15 16 17 18 19
20 21 22 23 24 25 26
27 28 29 30 31 1984
1984 JUNE

WATCH, a pocket instrument for measuring time, excited into action by a steel spring, coiled up, and acting by various ingenious contrivances. The spring is in a brass box, called the barrel, and combined with a pyramidal fusee, on which a connecting chain is wound by the key. The spring being fastened at one end to the barrel, and at the other end to an arbor, or axle, unwinds off the fusee, turning it, and keeping the watch going, while the action accords by its various size with the varied energy of the spring. The force being thus produced, other wheels are put in motion, and time is exactly measured by the hands on the dial. Watches were invented about the year 1500, and the trade is much esteemed in all countries. The Swiss have carried it to the highest degree of perfection.

Wheels, &c. of a Watch.

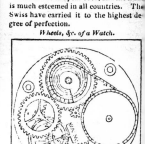

Spring of a Watch.

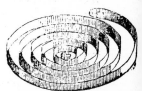

WATCHMAKER, an artist who arranges and puts together the wheels and parts of a watch, after they are cast and prepared by other artizans; and it is one of the most ingenious of all the mechanical arts.

TIME, the measure of motion, and of succession of thoughts, determined ... our globe by the phenomena of light ... darkness, and returning seasons of ... t and cold; the earth's motion round ... axis being a day, and its motion round ... sun a year; while the period from ... moon to new moon determines the ... nth. Longer periods are the moon's ... le of recurring phenomena in nine... n years; the earth's cycle, or the re... cession of the equinoctial points ... nd the ecliptic in 25,600 years; but ... revolution of the line of apsides is in ... 00 years. These are periods of dura... n relative to our globe and to us, but ... erent to different systems, and even to ... erent animals and states on the same ... net; yet all included in a totality ... simple duration, common to the whole ... verse, and which totality is necessarily ... hout measure, parts, beginning, or ... ing. Time is, therefore, an acquired ... nan mode of considering motions or ... nts, and our views can have no exact ... ation to its absolute properties as an ... versal totality.—*See Space.* Every ... ng is relative which is not infinite; ... d all are evolved and absorbed by the ... alities of infinite time and space. ... ne in the ancient mythology is drawn ... an old man, with a scythe and hour... ss.

e Rev. S. Barrow's *Popular ...ctionary of Facts and Knowledge*, 1827

TIME IN MUSIC, is of three lengths, as slow, or adagio, quick, allegro and

CHRISTOPHER RAWLENCE –
CLOCK TIME

Trains don't clacketty-clack like they used to. Not on the main lines, anyway. The time of journeys is no longer paced, metronome-like, by the regular sixty-six-foot intervals between the ends of each rail. In these days of welded track, it's more of a heavy steel whoosh, arbitrarily broken by the occasional clatter of points. The visual pulse of telegraph poles has gone as well, and with it, the mesmerising undulation of the wires.

A similar loss of visible and audible measure has happened to our clocks and watches. The reciprocal tick-tocking of balance wheel and escapement, the swinging of pendulums, which helped to create our sense of time in the act of measuring it, has been ousted by the silent micro division of time into a fractionalised digital blur. Comforting mechanical oscillations have given way to the high-frequency vibration of the quartz crystal. What we could once touch and tinker with is now, both literally and conceptually, hard to grasp.

At the official echelons of National Time Services (pip-pip-pip-pip-peeePP), even the rhythms of the quartz crystal are not accurate enough. Today's atomic clocks, which take 30,000 years to gain or lose a second, are monitored by the highly precise frequencies of the caesium atom.

The history of clocks is a technological quest for absolute accuracy and begs the question: accurate in relation to what? The answer might once have been: accurate in relation to the regular rotation of the earth, against which clocks were set. But today's clocks are more accurate than the earth and reveal its rotation to be slowing by one second a year. The seeds of an illusion seem to lie here: it's as if in harnessing the mysteries of the quartz crystal, or penetrating the heart of the caesium atom, we convince ourselves that science leads us to the true pulse of time itself. It's as if the precision of modern timekeeping lends subliminal weight to the idea, prevalent in the West for several centuries, that there is such a thing as THE time in the universe, and we can measure it accurately.

On the Leeds Executive (0750 King's Cross), clusters of businessmen are monitoring the hourly 'peep-peeps' on their digital jewellery. Some check the time against familiar landmarks along the track — when the train passes that strangely isolated church it should be close on 8.30 if we're running to time. There are appointments to be kept and deals to be done.

The mechanical clock first appeared in the churches and monasteries of thirteenth-century Europe in answer to a recently developed need to indicate the divisions of the day more precisely. The calling of the clergy was prayer — as much of it as possible for the salvation of everyone. St Paul had even advocated praying all the time, but early Christians had a living to make like everyone else, so set times were recommended.

In the sixth century, St Benedict had formalised these times. 'The Rule of St Benedict' set in motion one of those periodic shake-ups of Christianity designed to return the flagging faith to first principles. It contained the blueprint for the organisation and principles of the infant monastic system, in which time and timing were of central importance:

'The prophet saith: seven times a day have I given praise to thee. We shall observe this sacred number of seven if we fulfil the duties of our service in the hours of Lauds (3.00 a.m.), Prime (6.00 a.m.), Terce (9.00 a.m.), Sext (Noon), None (3.00 p.m.), Vespers and Compline . . .'

Not only prayer but everything else was timed in this regime where idleness was deemed 'the enemy of the soul':

'The brethren, therefore, must be occupied at the stated hours in manual labour, and again at others in sacred reading . . . they shall rise at the eighth hour of the night . . . with digestion completed . . . from Easter until September 14th, they shall start work in the morning at the first hour until about the fourth . . . let it be so arranged that there may be a very short interval after Matins, in which the brethren may go out for the necessities of nature, to be followed at once by Lauds, which should be said at dawn . . . if anyone arrives late at the Night Office, let him do penance by public satisfaction . . .'

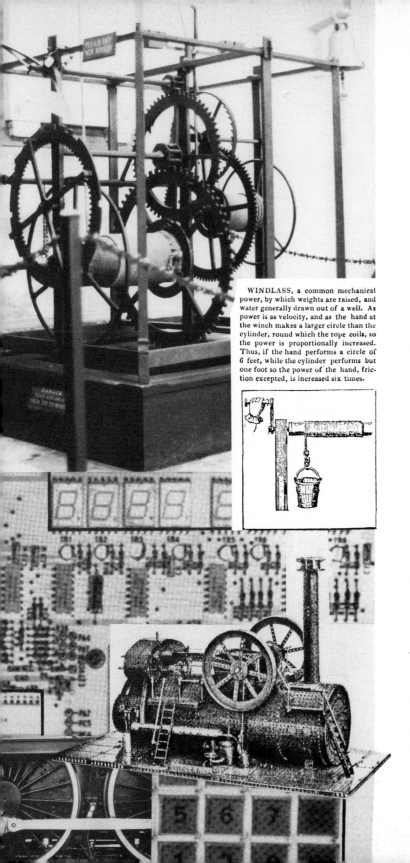

WINDLASS, a common mechanical power, by which weights are raised, and water generally drawn out of a well. As power is as velocity, and as the hand at the winch makes a larger circle than the cylinder, round which the rope coils, so the power is proportionally increased. Thus, if the hand performs a circle of 6 feet, while the cylinder performs but one foot so the power of the hand, friction excepted, is increased six times.

MIKE COOLEY – WORK AND TIME

The last two centuries have seen an extraordinary acceleration in the rate of technological change. It's almost as if this has made time itself appear to pass faster and faster.

Technological artefacts and equipment have a life cycle that's getting shorter and shorter all the time. Wheel transport, for example, existed in the same form for 2,000 years; Watt's steam engine was working for 102 years after it was built; in the 1930s, high capital equipment was built to be written off in thirty years; and today, the type of computer equipment I'm working with will be obsolete in three or four years' time.

In the past, people had skills and abilities which lasted them a lifetime, because the tools they were using and the products they were making would last a lifetime. But today, as tools become obsolete at an ever increasing rate, so too do the skills that people require to use them. They are trained to use a particular piece of equipment, but that knowledge is only valid for about two or three years.

This process has had three main stages of development. In the first stage we developed systems that could 'walk'. Human beings put part of themselves, in terms of strain energy, into the springs and weights of hand-wound clocks, an energy which was given up as the mechanism worked. There were also clockwork figures – elegant and complicated automatons that dated from medieval times. Much earlier than you'd think.

The next stage was the development of machines that could 'eat'. You no longer put strain energy into them by winding them up; you fed them instead – coal was shovelled into steam engines.

And today we've reached the stage where we have machines that can 'think': computers.

So we've been making machines that are images of ourselves. They can 'walk', 'eat' and 'think'. But the irony is that at the same time we've been transforming ourselves into the equivalent of machines. It's been a subtle dialectical process: in order to design a machine capable of doing the things we do, we have to think about how we do those things. The outcome has been a

33

Today, this punitive timetable seems like a gross parody of the grey heartlessness of our own industrial schedules. But St Benedict was very clear about its enlightening purpose:

'If there be some strictness of discipline, do not at once be dismayed and run away from the way of salvation, of which the entrance must needs be narrow . . . We shall share by patience in the sufferings of Christ, that we may deserve to be partakers also of his kingdom . . .'

There was a problem though. What's the use of a timetable if you can't implement it? As the French song implies, somebody had to 'sonnez le matine' in order to wake Frère Jacques. How was that person sure he was going to be awake himself in order to ring the bell? Furthermore, even if he was awake, how would he know what time it was?

St Benedict wasn't much help here: 'The indicating of the hour shall be the business of the Abbot. Let him either do it, or entrust the duty to such careful brother as would . . .'

As would what? Nod his way through the night, regularly turning an hour glass and marking time away with chalk marks on the wall, always in fear of falling asleep? Sustain a burning candle in a draught, losing count of the wax divisions as they burnt away? Cope with an iced-up water clock in winter? Consult a sundial on a cloudy day?

For six centuries they muddled through in this way, as the monastic system established itself, and the water clock (clepsydra) evolved into quite a sophisticated instrument. But by the thirteenth century, monasteries had become such complex centres of worship and economic activity, that they needed a more reliable way of telling time. The sheer scale of organisation demanded it.

The French word cloche is related to the English 'clock', and means 'bell'. As the language indicates, the first mechanical clocks rang bells rather than moved hands. They were mechanical alarm systems, powered by weights, that could be set for a certain time before tripping a bell. This would wake the 'careful brother', who would then toll the big bell, announcing this or that service. These early devices were not clocks as we think of them. They were more like darkroom clocks — their function was to indicate a time rather than tell the time.

Do you remember the last time you were late for school because the clock at home was slow? You can always have the correct time if you fit this shockproof clock to your handlebars.

CYCLE WATCH
Price
12'6
Post 4d.

THE ENEMY

Between here and there
The corridor journey
On two buses

To a clock
That rings
And a slot
With a name –
Mine

And the numbers
One two nine
These are mine
A card marks time

Among faces sad and disinterested
Amid wheels that hum to the strum
Of steel
Machines work the mind
To a gentle hypnosis

As you give all you can
Often more
To receive what is given
Which is less
Remember the others
And this life of no questions

And as you feel at your ease
And feel you belong
Beware of this ease
The prize of the blind
And the strong.

Mike Haywood – steelworker,
third-hand roller.

	1	2	3	4	5	6	7	8
Sun	6–2	OFF	10–6	2–10	6–2	OFF	10–6	2–10
Mon	2–10	6–2	OFF	10–6	2–10	6–2	OFF	10–6
Tue	2–10	6–2	OFF	10–6	2–10	6–2	OFF	10–6
Wed	10–6	2–10	6–2	OFF	10–6	2–10	6–2	OFF
Thu	10–6	2–10	6–2	OFF	10–6	2–10	6–2	OFF
Fri	OFF	10–6	2–10	6–2	OFF	10–6	2–10	6–2
Sat	OFF	10–6	2–10	6–2	OFF	10–6	2–10	6–2

A typical continental shift rota, as worked in many steel works. 6–2 = 6 a.m.–2 p.m.; 2–10 = 2 p.m.–10 p.m.; 10–6 = 10 p.m.–6 a.m. This rota covers an eight-week period.

In fact there wasn't a time as we think of it to tell. The length of the medieval European hour was irregular, like the Egyptian and Roman one. It was arrived at by dividing the time of daylight by twelve which meant that in winter an hour was shorter and in summer longer. This suited the rhythm of work on the land, where there was more to do in summer and less in winter. But it didn't really suit the evolving complexity of the monasteries, for whom regularity of schedule was very important. There was a need for the standard hour.

The essence of the mechanical clock is the creation of a regular beat through the controlled release of energy. The division of time into units of standard length is intrinsic to it, unlike the sundial or the water clock which measure time continuously. Any irregularity in the beat of a mechanical clock amounts to error.

It was a simple and logical step when the clock evolved from a resettable alarm device into a mechanism that meted out standard hours, give or take five minutes. It's tempting to ask which came first — the standard hour or the clock that measures it passing? Perhaps the nature of the mechanical clock encourages us to abstract time from nature itself. Sixty-minute hours weren't waiting to be measured; they were the creation of time measurement.

The mechanical clock first appeared to solve a problem of religious organisation, but soon there was a secular demand as well. If you were a prince or a

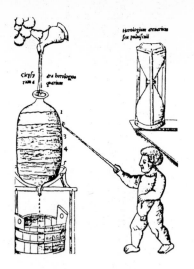

duke in one of the numerous medieval courts, what better way of flaunting your authority and prestige than by owning one of these magical machines? By possessing an object that appeared to harness hours, you might give the impression that you held sway over time.

But the most rapidly expanding market for clocks was the market-place itself, where the economic expansion of the late Middle Ages had changed quiet villages into thriving centres of commerce. A complex division of labour was emerging in these new towns, comprising buyers, sellers, artisans, entrepreneurs, labourers, craftsmen and even a new breed of administrator to hold it all together.

Out of this social melting pot emerged a new class of merchants. With their extraordinary wealth the more canny among them quickly became a powerful new élite. This nascent bourgeoisie were very soon controlling the affairs of the new towns, and they used some of their money to buy clocks.

These new town clocks, which were put up in belfries above the market-places, served two purposes. They were a symbol of the town's self-esteem, but they also regulated the work process. The latter meant more than simply a means by which busy merchants could keep appointments — 'We'll meet under the clock at noon.' The bells of the clock were time signals to apprentices and to the waged labourers of the medieval textile industry. They clearly demarcated when to start and stop work, and when to take a dinner break. This didn't simply mean that employers could be sure of the time their employees worked. It worked the other way as well, preventing unscrupulous employers from shortening lunch hours and lengthening the working day. Now everyone could hear what time it was. In its infancy, then, the mechanical clock was a potential arbitrator of conflicts about work and time.

'The clock, not the steam engine, is the key machine of the modern industrial age. At the very beginning of modern technics (in the thirteenth century) appeared prophetically the accurate automatic machine, which only after centuries of further effort was also to prove the final consummation of this technics in every department of industrial activity.'

Lewis Mumford: Technics and Civilisation

technology which tries to reduce our actions to equivalent machine-like motions. Today we have the idea that you can reduce the brain to some kind of computer mechanism.

How have you personally noticed the change in the nature of work in the drawing office?

In the 1950s we had time to enjoy our work, and our drawings were almost art forms in themselves. We'd take pride, if we drew a gear wheel, in ensuring that all the teeth were on it and drawn to diminishing pitch, so that we almost gave the sense that it was moving. You could say that we embellished our drawings.

But gradually, towards the end of the 1950s, management began to look at what they called the 'noise to signal ratio', which they started saying was unacceptable. This meant that the craft of our drawing took too much time and wasn't necessary to give the main signal.

They started producing charts which showed what they called rational drawing and design methods. This meant that instead of drawing the threads on a bolt, you drew two dotted lines indicating that the bolt should be there.

Then they started measuring the amount of time people took to do a drawing. By the mid-1960s they were even timing how long it took to sharpen a pencil. They might come into the office every twenty minutes to see how many were drawing, how many talking, how many were looking at books or were on the phone. And an efficient office came to be one in which the actual drawing activity predominated.

This seemed absurd because often the most creative ideas would come through just sitting and gazing, or apparently dozing, or chatting to a colleague. This creative time was regarded as an inefficiency; what you were supposed to do was consult the standard design manuals.

So a highly creative activity came to be one which was timed. And this was a prerequisite for the gradual integration of the drawing office into computer-based systems.

In my view, as a person draws an object they're internalising all kinds of things about it. It's the next best thing to actually making it.

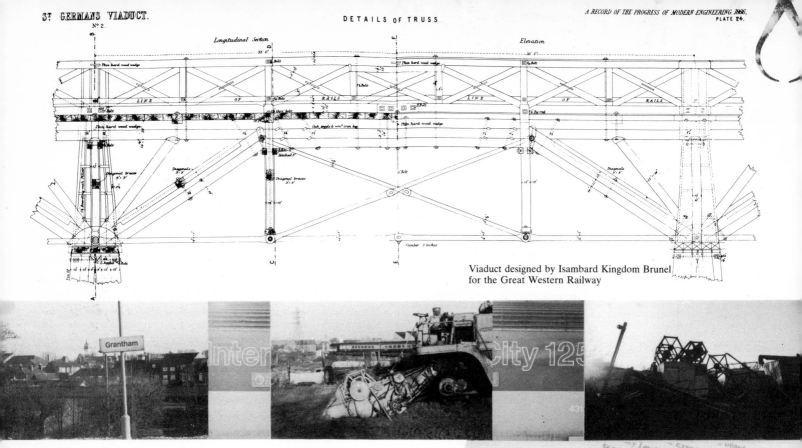

Viaduct designed by Isambard Kingdom Brunel
for the Great Western Railway

The Great Northern Railway is not as straight as Brunel's Great Western, but it's straight enough for trains to keep up 125 m.p.h. much of the way. Grantham is one place where they have to reduce speed. Approaching the town from the south, the train comes out of a cutting, and there's a touch on the brakes as it passes a combine harvester cemetery, before entering the curve. Then at a modest 100 m.p.h., it honk-honks through the station before skirting the town in a delicate arc.

If you look to your right across the rooftops of this market town, beyond the Isaac Newton supermarket you'll see the town hall clock presiding over the community. What you won't see, however, as the diesels whistle up their revs to pick up speed out of the curve, is the massive bronze statue of Isaac Newton down below the clock.

Speeding through Newton's home town, I watched a student swotting 'O' Level Physics, and slowly recalled that Newton's thought three centuries ago was underwriting the capacity of that steel mass to stay on the rails and run to time. It's a basic requirement of designing a train to know about the forces that will accelerate and brake it: power to weight ratios; friction between wheels and track; wind resistance and so on — all of which take Newton's laws of motion for granted.

In order to describe motion in a precise way, Newton found he needed to express time mathematically. In his equations of motion — $v = u + at$ and $s = ut + \frac{1}{2}at^2$ — 't' (time) took on an abstract objective reality for the first time. This is how Newton expressed it:

'Absolute, True and Mathematical Time, of itself and from its own nature, flows equably, without relation to anything external and by another name is called duration . . . all motions may be accelerated or retarded but the flowing of Absolute Time is not liable to change . . .'

In other words, time flowed in the same direction, in a straight line, all over the universe. The time on earth was the same as the time on the most distant galaxy, as if there was a universal clock.

There were draughtsmen I knew whose drawings you could recognise from the other side of the office because of the beautiful printing or cross-hatching. The computer has now eliminated all this and depersonalised it. By the late 1980s, drawing as a form of communication between designer and manufacturer will have gone. We should reflect on what this might mean for human beings: since earliest times, when people drew on cave walls, we've tried to project ourselves through the images we produced. This is now disappearing.

Computer-aided design is based on a 'menu' of standard elements which you call up. The menu is the same for everybody; you can't change it; and it serves up its elements very, very fast. So not only is the creativity of the designer severely limited but the time taken to do a drawing is, in some cases, reduced to one-twentieth of the time taken by a fast draughtsman.

The computer sets up a frantic tempo because it can handle the quantitative number-crunching elements so fast. It paces the worker: we've identified systems where the decision-making rate of the designer can be forced up by 1,800 per cent. This puts immense pressure on what little creative role the designer has left to play: the rate at which the computer demands decisions reduces his creativity by 30 per cent in the first hour, 80 per cent in the second hour, and thereafter he's just shattered.

So the human being becomes a bottle-neck in the work process.

I've seen how people using visual display units (VDUs) are made conscious of time. Some are programmed so that if you don't handle the data on the screen within seventeen seconds, it disappears. Medical people have noticed that operators show all the recognised signs of stress as the time approaches for the image to disappear. From the eleventh second they begin to perspire, then the heart rate goes up. Consequently they experience enormous fatigue.

The computer can also tell the rate at which you use it,

by recording how often you draw or call up something on the screen. And it's now becoming possible to distinguish between active and passive reception of information: changes in pupil diameter can indicate the difference between looking and seeing, between hearing and listening. So management will be able to know whether you're taking an interest in the image on the screen or not.

And now the amazing discovery has been made that as you get older so you get slower, something I knew as a child of five when I looked at my grandparents.

The 'peak performance age' for people of particular specialisations is being worked out by a whole range of researchers. People from different age groups are made to sit in front of a VDU solving problems of different complexity. The speed at which they do something with the data is measured, from which peak performance age can be calculated.

A series of experiments in the U.S.A. has shown that, as a structural engineer, my peak performance age is thirty-four, which means I'm thirteen years over the hill! We're apparently more durable than mathematicians (twenty-three) or theoretical physicists whose peak is slightly later at twenty-seven.

After the peak there should be what is called a 'careers plateau', and thereafter a 'careers de-escalation', which

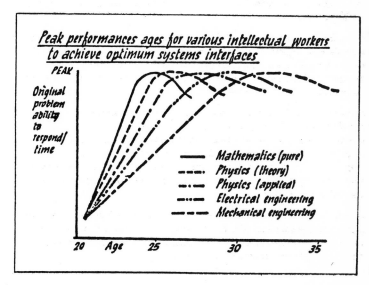

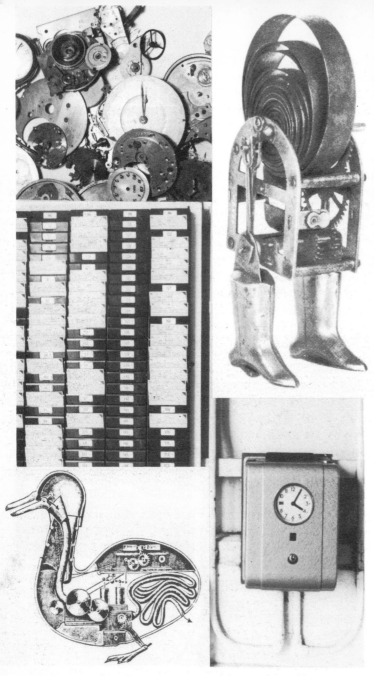

The idea of a universal clock meting out the one time of the universe is a facet of European seventeenth- and eighteenth-century philosophy in which clockwork itself had become a metaphor for the nature of the universe. The mechanics of clockwork — in automata and orreries as much as in clocks themselves — seemed to embody a perfection of cause and effect that was a microcosm of God's perfection and the workings of the universe. Philosophers and scientists expressed this aesthetic most clearly:

'The universe is a great piece of clockwork' (Boyle); 'The universe is not similar to a divine living being, but is similar to a clock' (Kepler); and Rousseau saw God as the divine clockmaker of a clockwork universe.

Descartes, who provided the philosophical framework for Newton, extended the image of clockwork machinery from the structure of the universe to life itself. In suggesting a separation between mind and matter, he elevated thought above the clockwork reality of plants, animals and even the human body: 'I consider the human body as a machine . . . my thought . . . compares a sick man and an ill-made clock, with my idea of a healthy man and a well-made clock.'

Even more explicitly, Descartes wrote: 'We see clocks, artificial fountains, mills and other similar machines which, though merely man-made, have none the less the power to move by themselves . . . I do not recognise any difference between the machines made by craftsmen and the various bodies that nature alone composes . . .'

This kind of mechanical causality was to be at the heart of the Industrial Revolution. Steam engines and power looms were to be working and profitable proof of Newton's laws of motion. But the 't' of his Absolute Time wasn't simply a function in the dynamics of machinery. In a strange way it sanctioned the authority of the factory clock, permeating the labour process, which it chopped into hours and priced. 'Newtonian-Cartesian' thought provided the foundation for the science and technology we have today. By a kind of cultural osmosis it legitimated the worst excesses of industrialisation where machine production mechanised the people who worked in it. Taylor's Scientific Management of the early twentieth century, enshrined this Enlightment thinking on the shopfloor.

'In the Newtonian view, God created in the beginning, material particles, the forces between them, and the fundamental laws of motion. In this way the whole universe was set in motion, and it has continued to run ever since, like a machine, governed by immutable laws . . . all that happened had a definite cause and gave rise to a definite effect, and the future of any part of the system could — in principle — be predicted with absolute certainty if its state at any time was known in all details.'   Fritjof Capra: The Turning Point

means a gradual phasing out: if you look in the computing department of any multinational corporation you'll find very few people over the age of thirty.

With word processors – the so-called lower levels of intellectual work – work rate is already being measured. At the end of a day management can have a print-out that tells them how often and how many keys were depressed during the day and how many mistakes were made. While you were using the delete button because you'd thought of a more elegant sentence, the machine would be taking it as a measure of your errors.

There was a conference in London recently at which one of the leading word processor manufacturers said: 'Your secretary will be required to be more productive. There will be no more walking, talking, thinking or dreaming.'

Yet these are the most precious attributes of human beings. The same manufacturer went on to say that the new technology will provide a basis for introducing Taylorism into the field of intellectual work.

What is Taylorism?

Taylor was an American industrial engineer working at the turn of the century. He believed that it was possible to organise industry in a much more rational and efficient way, by eliminating movements and activities in the work process that were superfluous to the end result.

In his book, *Scientific Management*, the famous example he gives is shovelling pig-iron. He worked out the optimum way of shovelling it: the minimum number of motions with the maximum amount in each shovel-load in relation to the capacities of the worker.

The idea was that it would be both to the advantage of management and workers. Management would get higher productivity and a higher rate of profit; and through producing more workers would be compensated through higher earnings.

Taylor thought he was doing something helpful for both parties. But frequently something that starts out as a benevolent idea is transformed into its opposite. Taylor ignored the fact that there are conflicts of interest in industry, and that there are human requirements that transcend the requirements of the machine.

Taylor's ideas were first implemented on a large scale in the motor industry. At Ford's, for example, the worker no longer had the freedom to walk around the static vehicle as he assembled its different parts. Instead, he was restricted to a repetitive task in a fixed position as vehicle after vehicle passed him on the production line.

INAUGURATION OF THE STATUE OF SIR ISAAC NEWTON AT GRANTHAM.—SEE SUPPLEMENT, PAGE 315.

When assorted worthies gathered at St Peter's Hill in Grantham on October 2nd, 1858, for the inauguration of Newton's statue, they were possibly doing no more than expressing their pride in the local boy made good. But the coverage given the event by the Illustrated London News *suggests that it meant much more. It's as if they were celebrating Newton as the guiding spirit of Britain's industrial progress. An influence that Blake certainly recognised and passionately damned in his work.*

42

THE INAUGURATION OF THE STATUE OF SIR ISAAC NEWTON AT GRANTHAM.

THE statue erected to Sir Isaac Newton at Grantham, Lincolnshire—near which town he was born, and in which he received the rudiments of his education—was inaugurated on Tuesday week, as duly recorded in the last number of this Journal, with great ceremony, and in the midst of a vast concourse of persons, including many men of science from all parts of the country. On the first page of our present Number we have illustrated this interesting Inaugural Ceremony, and on the present page we give an Engraving of the Statue.

To the particulars of the inauguration given in this Journal last week we add the following account of the procession. At one o'clock the noblemen and gentlemen taking part in the procession met at the Grammar-school, and proceeded thence to St. Peter's-hill in the following order :—

Escort of Military.
The Band of the Royal South Lincolnshire Militia.
Town Flag.   Crier.   Town Flag.
Policemen.   Chief Constable.   Policemen.
Mace Bearer.   Mace Bearer.   Mace Bearer.
Ex-Mayor.   The Mayor.   The Recorder.   Town Clerk.
The Vicar and Parochial Clergy of the Borough.
The Aldermen and Borough Magistrates.
The Town Council—three abreast.
The Boys of the Grammar-school—four abreast.
Head School Boy, carrying the "Principia."
Second Boy, carrying the Reflecting Telescope invented by Newton.
Third Boy, carrying Newton's Prism.
The Masters of the Grammar School.
The Lord Bishop of the Diocese.
Dr. Whewell.   The Right Hon. Lord Brougham.   Professor Owen.
The Committee of Selection.
Major-General the Hon. Sir E. Cust, K.C.H.   Sir G. Welby, Bart.
The Sculptor.   The Founders,
W. Theed, Esq.   Messrs. Robinson and Co.
The Secretaries.
The Members of the Committee—three abreast.
Gentlemen attending by invitation—three abreast.

The route was by Church-terrace, Vine-street, and High-street. The appearance which the procession presented, banners flying and band playing as it approached the place of the statue, was very imposing. On their arrival Lord Brougham took his seat on a chair which was said to have belonged to Sir Isaac, and was loudly cheered. Immediately behind him was seated the Mayor of Grantham, Mr. Ostler, and the Bishop of Lincoln. On the seats around were Sir J. Trollope M.P., Mr. Milnes, M.P., Dr. Whewell, Professor Owen, Sir J. Rennie, Sir Benjamin Brodie, Lord A. Compton, Dr. Latham, and a numerous assemblage of ladies and gentlemen. When all were seated, the covering, which till now hid the statue from view, was withdrawn, and a chorus of cheers burst from the assemblage. The noble statue stood out in the bright sunshine in all its elegant proportions, and its character and expression were the theme of general admiration.

When the cheering attendant upon the withdrawal of the covering from the statue had subsided, Lord Brougham delivered a magnificent oration, of which we were unable last week to give more than the concluding portion. His Lordship spoke as follows :—

We are this day assembled to commemorate him of whom the consent of nations has declared that that man is chargeable with nothing like a follower's exaggeration of local partiality which pronounces the name of Newton as that of the greatest genius ever bestowed, by the bounty of Providence, for instructing mankind on the frame of the universe, and the laws by which it is governed—(the noble Lord was here overpowered by emotion, and paused : in a few seconds he proceeded)—

Whose genius dimmed all other men's as far
As does the midday sun the midnight star.

But, though scaling these lofty heights is hopeless, yet is there some use and much gratification in contemplating by what steps he ascended. Tracing his course of action may help others to gain the lower eminences lying within their reach ; while admiration excited and curiosity satisfied are frames of mind both wholesome and pleasing. Nothing new, it is true, can be given in narrative ; hardly anything in reflection ; less still, perhaps, in comment or illustration ; but it is well to assemble in one view various parts of the vast subject, with the surrounding circumstances, whether accidental or intrinsic, and to mark in passing the misconception raised by individual ignorance or national prejudice which the historian of science occasionally finds crossing his path. The remark is common and is obvious, that the genius of Newton did not manifest itself at a very early age ; his

STATUE OF SIR ISAAC NEWTON, INAUGURATED LAST WEEK AT GRANTHAM.

the face of the science, effecting a revolution philosophy connected with it. Before 1661 he had committed to writing the metho years of age he had discovered the law of tion of celestial dynamics, the science cr had elapsed he added to his discoveries of light. So brilliant a course of disc and reconstructing analytical, astro defies belief. The statement could o the incontestable evidence that pro these doctrines gained the universe clearly understood, and their orig question. The limited nature of his ever reaching at once the ut Survey the whole circle of th progress in each—you find th law of gradual progress con equally governs. Again degrees, from its first ru their lords' courts, and money—the great disc politics has been effect of any extent to enjo be established, comb statesmen and writ wholly impossible its inhabitants, sudden and rapi prepared by th several succes have had, u pierced the The arts a forerunne rule. Th had been and the of their tactics. and New cised on step whi possibly "Princi perhaps those her ledge, aw naturally o great mechan as it is powerful one discovery o the degree in after, also liv that in this re bounds of k discoveries i feats of sci character. trating t undisput might be he coul the ink from t them the c men fert su al

faculties were not, like those of some great and many ordinary individual precociously developed. His earliest history is involved in some obscurit and the most celebrated of men has, in this particular, been compared the most celebrated of rivers, the Nile—as if the course of both in feebler state had been concealed from mortal eyes. We have it, how well ascertained that within four years—between the age of eighteen twenty-two—he had begun to study mathematical science, and had his place amongst its greatest masters, learnt for the first time the el of geometry and analysis, and discovered calculus which entirely

The train has distinctly slowed on the stretch between Grantham and Newark. It doesn't quite seem to have the power to sustain 125 m.p.h. on the shallow uphill gradient, which must mean a mechanical problem. An atmosphere of irritable anxiety has begun to permeate the carriage as businessmen, checking their watches, realise they are going to be late. I can hear the conversation at the next table and they're blaming the trade union pig-headedness of the driver as if it were a deliberate go-slow. They think he is working to rule in protest against the recently introduced flexible rosters which the driver's union bitterly opposes.

The Industrial Revolution, which began in the late eighteenth century and arguably is not yet complete, imposed a new discipline of time upon its first generation of workers. Many early factory 'hands' came from working on the land where they had reckoned time by the movement of the sun, and measured it by the time it took to do something — hoe a row, for example, or milk a cow. Time sense was task-oriented — the closest present-day analogy might be the way we think of the time it takes to boil an egg. It wasn't that they were unaware of hours in eighteenth-century rural England — they could hear the parish clock — but they didn't need to measure time precisely or worry about punctuality. They were not clock-conscious.

The new factory system changed all this. It was based on the co-ordination of large numbers of workers in the same place for the same duration — you had to be there on time, for a long time. But it was more than time in the form of the factory clock that imposed the new time discipline on the workforce. The

machinery itself demanded dehumanised clockwork responses from those who operated it. It was imbued with the clockwork mentality. Indeed, most of the early machines were made by clockmakers, who were particularly well qualified because of the nature of their work. Listen to the angry clockmakers who opposed William Pitt's planned tax on watches in the 1790s to finance the Napoleonic Wars:

'The cotton and woollen manufactories are entirely indebted for the state of perfection to which machinery used therein is now brought, to clock and watchmakers, great numbers of whom have, for several years past . . . been employed in investigating, constructing, as well as superintending such machinery . . .'

Ex-farm labourers were not the only new factory fodder. There were also artisans and craftsmen whose skills had been made redundant by the new machinery. Many of these people — weavers and spinners, for example — had been used to working as a family unit in a cottage industry. Supported by a plot of vegetables and a few animals, they produced at their own rate. This was no golden age — bad times were very bad — but in good times they might only need to work four days in a week. Through owning their means of production — a handloom for example — they were in command of their own time and could regulate their work rate. Until well into the nineteenth century the practice of Saint Monday persisted in the North and Midlands of the country. By working hard from Tuesday to Friday, weavers, cobblers and others found they could take a long weekend. In France this same custom was characterised by the saying 'Sunday is for Worship, Monday is for Love.'

The factory clock and the clockwork rhythm of machinery were a constant reminder to this early generation of factory workers about their changed experience of time. No longer owning their means of production, they now had only their labour time to sell. Time became the central issue — how many hours you worked, the rate you got paid for them, and their quality. At first,

Pieces	Reach into Tote Pan, Pick up Piece, Place in fixture, close fixture HAND LEAVES FIXTURE		Turn the fixture ¼ turn, Place piece in position under Drill Start DRILL TOUCHES WORK-PIECE		Drill Hole DRILL CHUCK HITS FIXTURE		Drill up, Turn fixture, Unlock fixture, Remove piece DROP FINISHED PIECE INTO CHUTE		OPERATION

DRILL .123" HOLE INTO LIGHT ALLOY STRIP

Study commenced	9.00 a.m.	21/9/67
Study completed	9.20 a.m.	
Lapsed Time	20 mins.	
Effective Time		
Ineffective Time (D.N.A.)	9.50 mins.	
Synchronisation Time Beginning	2.86 mins.	} 20 mins.
and end of study	7.64 mins.	

	R	T	R	T	R	T	R	T		FOREIGN ELEMENTS AND DELAYS
1	5.16	.16	5.26	.10	5.66	A (.40)	5.88	.22	A	Several Unnecessary motions D.N.A. •
2	6.08	.20	6.18	.10	6.42	.24	6.62	.20	B	Drill Hit Hard Spot–will be eliminated D.N.A. •
3	6.78	.16	6.90	.12	7.12	.22	7.34	.22	C	Pieces Stick...
4	7.54	.20								

SELECTION OF RELAXATION ALLOWANCES

Dept. ___ Kitchen ___ Date 1-1-68

Operation ___ Making a pot of tea and laying a tray

Sex F.

TASK OR ELEMENT	Personal Needs	Minimum Fatigue	Standing	Abnormal Position	Weightlifting or Use of Force	Air Conditions	Light Conditions	Visual Strain	Aural Strain	Mental Strain	Monotony Mental	Monotony Physical	TOTAL
	7%	4%	4%	2%									
A. Walk to sink, get and fill kettle (5 lbs.)	7	4	4		1								16
B. Walk to cooker, light and put on kettle	7	4	4		1								16
C. Get 2 cups etc. and put on tray	7	4	4								1		16
D. Get milk and sugar and put on tray	7	4	4										15
E. Get tea-pot and walk to cooker	7	4	4										15
F. ... pot, fill tea-pot and put on tray	7	4	4								1		16

Give allowances for these factors only if conditions cannot be improved

.158	.112	.250	.238
95	90	M/C 100	80
.150	.100	.250	.190

Average Time per piece =	.758 min.
Normalised Time per piece =	.6800 min.
+Percentage allow. 15% (per./fat./cont.) =	.1034 min.
TOTAL TIME PER PIECE =	.7934 min.
PIECES PER HOUR	75

From 'A Guide to Time Study' issued by Executive Council of the Engineering Union

On one occasion Henry Ford even said that the person who puts on the screw doesn't screw on the nut. Labour was even divided down that far.

There's a famous trade union agreement in the car industry, whereby the total time allowed the worker for the work cycle is 32.4 minutes.

The time of a work cycle, according to the work study people, is the time taken from the beginning to the end of a given task. This could be putting the wheels on a car, or spot welding certain components together.

But the elements that make up this particular agreement are unbelievable: 'trips to the lavatory ... 1.62 minutes (not 1.6 or 1.7 because this is computer precise!) ... recovery from fatigue ... 1.2 minutes ... sitting down after standing too long ... 65 seconds ... monotony ... 32 seconds ...' and so the grotesque litany goes on.

Now some technologist had the arrogance to do this to another human being. And we wonder why there are strikes in this country!

Do you think there's a sense in which all strikes revolve round an issue of time?

My colleagues in the Engineering and Transport Unions have often pointed out that although strikes are usually sparked off by a narrow economic issue, behind it is the desire to demonstrate that the human being still has some control over the production line. In other words, they can stop it.

The alienation that comes from machine-paced production was dramatically shown at the General Motors plant at Lordstown in the U.S.A., where the workers kicked in some of the car windscreens. When asked why they'd done it, one said that if he couldn't improve the quality of the bloody thing at least he could make it worse. This was a pathetic attempt to show that human beings have some control over the means of production, even in a negative way.

Taylor himself said: 'In my system the workman is told precisely what to do and how he is to do it. Any improvement he makes on the instructions given him is disastrous to success.'

But if you reduce everything down to little depraved segments of work and allocate precise times to each of them, you turn human beings into machines. And even if you do pay them more, in my view this results in an enormous cost in terms of human alienation, dissatisfaction and unhappiness.

as is well known, the 'masters' had the upper hand. A local minister wrote of a Calderdale mill:

'If there was one place in England that needed legislative interference, it was this place, for they work fifteen and sixteen hours a day frequently, and sometimes all night. Oh! it is a murderous system, and the mill owners are the pest and disgrace of society . . .'

Few workers owned their own watches in these early years of the century. Prohibitive cost was not the only reason, as a Dundee textile worker recalled in his old age: 'In reality there were no regular hours. The masters and managers did with us as they liked. The clocks in the factories were put forwards and backwards, morn and night. Instead of being instruments for the measurement of time, they were used as cloaks for cheatery and oppression. A workman was afraid to carry a watch, as it was no uncommon event to dismiss anyone who presumed to know too much about the science of horology.'

But the nature of the new work discipline was fought against from the outset. The formation of trade unions, which rapidly became illegal, was a direct response by workers to the qualitative abuse of their time by the new system and its employers. When the Luddites smashed powerlooms in the early years of the nineteenth century, they were fighting the force of industrial time as they saw it embodied in machinery. It wasn't mechanisation as such that they opposed, but the inhuman way in which it was used.

As the century progressed and industrial capitalism consolidated itself, the terrain of this struggle over time shifted. Instead of fighting the very nature of industrial time, and therefore the system, the trade union movement emerged from illegality, fighting to get its membership the best possible deal within the system. The fight was no longer against the new industrial time, but about it. The new situation became apparent in the struggle for a shorter working day. The Ten Hour Act of 1847 was the outcome of long and widespread agitation and defined a statutory limit for the first time. This was followed by the Nine Hour Movement, and then an Eight . . .

I can see the guard approaching down the aisle of the carriage. He's in a hurry and won't stop to answer the questions of angry travellers as to why the train is running so late. I suspect he's on his way to the back of the train to make an announcement. His jacket is open and as he bustles past me I can see a watch chain leading from a buttonhole in his waistcoat to the 'Guard's Watch' in a pocket.

The reluctant acceptance of the new temporal order by the working class is reflected in the vast increase in watch ownership that took place during the nineteenth century. Around 1800 a watch would have been hand-crafted and cost several guineas. On the whole, only the gentry, factory owners, farmers and tradesmen would have been able to afford one. But by the 1850s, the

Waltham Watch Company of Boston, for example, could produce the same watch for a pound in vast quantities. Most Manchester operatives are said to have possessed a watch by the middle of the century. In the Bradford area Watch Clubs were formed to facilitate ownership. Members would meet once a week at their local pub, subscribing sixpence a time for a year. As soon as the pool totalled £1 5s. a silver pocket watch was purchased and raffled. This process was repeated weekly until all members had received their watch.

In a society increasingly dominated by railway timetables and work schedules, a watch was very useful. For those who worked on the railway it was indispensable. But a watch also conferred status: your time might no longer be your own, but at least you could own the means of telling it. Perhaps, at a deeper level, the more people felt they had lost control of their time, the more they needed to display the illusion of possessing it.

FATTORINI'S
FOURTH WATCH CLUB
FOR CARLET

Daily Mail, Monday, April 19, 1

As the third Club has now expire satisfaction, it is the wish of several of the mence a FOURTH ONE.

FIRST MEE THE

On FRIDAY Evening, Decembe WILL TAKE PLACE
AT EIGHT O'CLOCK,
At Mrs. ELLISON'S, Swan

When a committee will be chosen for the management of the Club; a Jewellery, &c., will be shown.

Persons desirous of obtaining information respecting the principles of the Club derived by the members, are requested to attend.

LIST OF ARTICLES SUPPLIED TO TH
Gold Watches	£3 to £25	Tea Spoons
Silver Watches		
Gold Albert Chains	£1 1s. to	Spoons
Clocks of all kinds	£1 10s.	£15 Des
Wedding Rings	3	adi Bags
Cruet Frames	7	Work Boxe
Knives and Forks	from 3 6 half d	g Desks
Britannia Metal Tea Pots	8	IR BRUSHES
Britannia Metal Coffee Pots	5/- to 14/-	ET GOODS of ev
		SEWING MACHIN

Printed at Edmondson and

Daily Mail COMMENT

Last straw or first rumble?

HERE we go again. A BL plant is pro ducing ultimatums instead of cars. Men at Cowley have been given th choice : Stay on strike after tomorro and you are sacked.

Management wants to abolish th custom whereby the men end wor morning and afternoon, three minut early.

Shop stewards say that the manageme is becoming increasingly 'autho tarian' and this is the last str Management retorts that, as the fi real hint of economic revival, the st stewards and the unions are trying grab the chance to reassert some their lost power. Very probably, the truth in both versions.

BL management has learned to tough. It has had to. For the c pany could not have survived oth wise.

Now BL has, in the Metro and Mae (produced at Cowley), good models that sell. It has also re tered dramatic improvements efficiency.

But competition in car salerooms never been more fierce. BL has even better. Every minute on assembly line does count.

A relentlessly repetitive way to e living. There is, unfortunatel other way to make cars for the market. BL cannot opt out. C cannot opt out.

The six-minute washing time st irresponsible when the West Mi and, indeed, the Western wo sadly over-supplied with redu car workers, who have 24 hours seven days a week to wash an clean.

Isn't one option to embrace automation, and welcome robotics on the shopfloor, as something that will liberate people from degrading work?

Liberate them to do what? Stand in the dole queue?

—*The Eiffel Tower in course of Construction.*

Rightly or wrongly I believe that work is central to the very essence of human beings. And I don't mean by this the grotesque alienated jobs at Ford's, but work in the historical sense which used to link hand and brain in a meaningful creative process. Human beings learn and develop through this, by handling uncertainty, organising and selecting materials. We interpret the world by acting on it, and develop a whole range of skills through this interaction. We develop massive bands of knowledge in this way. A skilled worker, for example, turning something on a lathe, knows from hand pressure whether it will stay in the chuck, knows how deep he can cut, knows the material he's working with and knows if something's the right shape or form by looking at it. And these judgments will be right every time, because they're based on the tacit knowledge of a lifetime of working on the real world. If you deny people this, you deny them part of their self-image.

Now I have to admit that a lot of work has got so

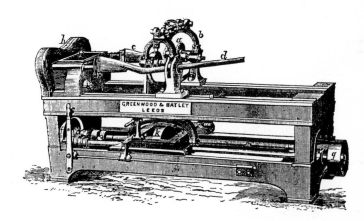

The Blanchard Lathe.

Work, for the Night is Coming.

(WORK SONG.)

LOWELL MASON.

This was the kind of moral compunction experienced by many working-class people in the nineteenth century and it had the effect of turning them into their own taskmasters. The moral blackmail of Christianity nurtured the psychological acceptance of the new industrial time. The price of pleasure in this life — meaning lightness, loving, dancing, trifling, singing, too much sleep, overeating, cardplaying . . . in fact just about everything that makes life worth living — was your salvation. The consequence of not working hard enough in this life was damnation in the next.

Behind the veil of a religion that exhorted the Imitation of Christ in sacrificing one's life for him, lay the sordid daily reality of selling that life for wages. Benjamin Franklin summed up this reality in the context of pioneering America:

'Remember that Time is Money . . . he that idly loses five shillings worth of time, loses five shillings, and might as prudently throw five shillings into the sea . . .'

With the Industrial Revolution, time became currency. Instead of being 'passed', it was now 'spent', 'invested', 'wasted', 'saved' and 'budgeted'.

How did the acceptance of such a deep restructuring of human time happen so quickly? Force of circumstance and sheer economic necessity played a major part. If you were starving and faced with the poor house, the factory was the lesser of two evils. There was also physical coercion — the regular use of police and army to break strikes and crush demonstrations. But the inner force of religion played its part as well.

The idea of time as linear does not simply stem from Newton's thought. It is inherent in our sense of history, which perhaps found its first expression in the linear narrative of the Bible, where time is laid out in a line from Genesis to Revelation. Christianity sees the individual human life as a microcosm of the greater Biblical drama — the line of one's lifetime between birth (creation) and death (salvation through return to heaven). The Puritans of seventeenth-century England enhanced this linear conception of life time by putting great moral emphasis on it: 'Remember still that the Time of this short uncertain life is all that you shall ever have, for your preparation for your endless life. When this is spent, whether well or ill, you shall have no more.' The prevailing Christian ethos in nineteenth-century Britain was founded on this Puritan outlook. As Dr Ure, an apologist for the factory system, wrote in his Philosophy of Manufacture (1835), 'the first great lesson . . . [was that] . . . man must expect his chief happiness not in the present, but in a future state.' In a life of appallingly delayed gratification, work was to reign supreme as 'a pure act of virtue inspired by the love of a transcendent being'. Christ's sacrifice was the shining example of this punitive morality: 'It is the sacrifice that removes guilt of sin . . . it is the motive which removes love of sin . . . it is the pattern of obedience.'

Hans Arp: Notes from a Dada Diary 1932.

monsieur duval

man is a beautiful dream. man lives in the sagalike country of utopia where the thing-in-itself tap-dances with the categorical imperative. today's representative of man is only a tiny button on a giant senseless machine. nothing in man is any longer substantial. the safe-deposit vault replaces the may night. how sweetly and plaintively the nightingale sings down there while man is studying the stock-market. what a heady scent the lilac gives forth down there. man's head and reason are gelded, and are trained only in a certain kind of trickery. man's goal is money and every means of getting money is alright with him. men hack at each other like fighting cocks without ever once looking into that bottomless pit into which one day they will dwindle along with their damned swindle. to run faster to step wider to jump higher to hit harder that is what man pays the highest price for. the little folk song of time and space has been wiped out by the cerebral sponge. was there ever a bigger swine than the man who invented the expression time is money. time and space no longer exist for modern man. with a can of gasoline under his behind man whizzes faster and faster around the earth so that soon he will back again before he leaves. yesterday monsieur duval whizzed at three o'clock from paris to berlin and was back again at four. today monsieur duval whizzes three o'clock from paris to berlin and was back again at half past three. tomorrow monsieur duval will whiz at three o'clock from paris to berlin and will be again at three o'clock that is at the same time he leaves and day after tomorrow monsieur duval will be back before he leaves. nothing seems more ridiculous than broad clear living.
present-day man

grotesque that people are better off out of it. But why has it become so grotesque?

Skilled work on universal lathes, milling machines etc., used to be some of the most creative and fulfilling work on the shopfloor. This is now being replaced by computer-controlled machine tool systems. They've succeeded in de-skilling the work so much that one U.S.A. manufacturer of this equipment advocates that management select workers with a mental age of twelve to operate it. Had the aim been to create work for the mentally retarded, rather than cut wage costs, this would have been admirable.

My view is that we redesign our tools to make full use of human creativity and intelligence. Because in spite of all the robotic equipment we've got, compared with human intelligence these systems are trivial.

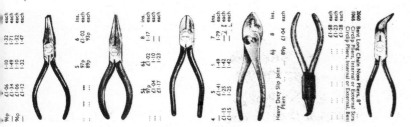

The Japanese would say that if you take account of the capability of machines to analyse their own faults, their intelligence in 'switching unit' terms is ten to the ninth. But by the same comparison, the intelligence of the human being is ten to the twenty-sixth: many millions of times greater, and it brings with it imagination, consciousness, will, ideology, humour ... all those things which make us unique and precious as human beings, but precisely all those things that modern management doesn't want.

Indeed there's one British manager who'd say these qualities represent bloodymindedness, yet it's precisely these attributes that make people capable of developing new systems.

But what are the practical alternatives? Are you suggesting we reverse these trends in technological development?

With a number of colleagues (at Manchester University) I've been working on what we describe as human-centred systems. This means that instead of the machine determining the rate at which humans react, systems are designed to respond to human intelligence, time and pacing.

Yes. This would be quite contrary to the main historical tendency, which is to absorb human intelligence into the machine, control the human being with the machine, and even look for human beings that are compatible with the machine.

We've been designing computer-based lathes which take account of the subjective, tacit, qualitative judgment of a worker. This means that he or she can stop to contemplate what they are doing: the worker paces the machine, rather than the other way round. We then get what we call a machine/human symbiosis, where the best of the human being is linked with the best of the machine.

But surely it's not realistic to make the space for taking your time in a modern industrial economy? And I'm not just talking about Western capitalism here. I imagine the same would apply to a Socialist economy like that of the U.S.S.R. In the end it seems to come down to the pressure to create material wealth, whatever society you're in.

Yes. Lenin advocated Taylorism in the Soviet Union and I think it was a terrible mistake. I'm not opposed to creating material wealth, but there are other ways of doing it.

Let's say the requirement is for people to have a car for twenty years. Well, they can buy five cars, each of which lasts four years, then falls apart. This is the way things are done now, frantically, on production lines that are so alienated that eventually you have to get robots to do the job.

The other way, of course, would be to design and build good-quality cars that would last for twenty to thirty years. This would be quite easy to do in technical terms, and the people that produced them would be doing interesting skilled work. The people who make Range Rovers, for example, see it as a good-quality vehicle, and so they work on it with a much higher level of commitment and motivation than on some of the more 'throwaway' products.

49

There's a mushy excuse crackling out over the P.A. The guard is regretting that we're running late, but it's due to 'technical difficulties'. I know what this means: we've lost a power car, which makes this the fifth engine failure I've noticed in the past month. A driver once explained it to me as what happens when aircraft designers and marine engineers try to design trains. The diesels fail so regularly because they're designed for boats, rather than hammering up and down 1,000 miles of track a day inside a train.

This means we'll be forty minutes late. We're limping up the slow line which still goes clacketty-clack while they let the Edinburgh train past on the fast line.

GREENWICH OBSERVATORY ILLUSTRATED

The time discipline imposed on working people by the factory system was complemented on a national scale by the growth of the railways.

Because of the earth's rotation there is a twenty-five minute difference between the time at Dover, for example, and the time at Falmouth. Until half-way through the last century every town in Britain kept to its own local time, setting their clocks when the sun crossed their local meridian at noon. But with the coming of the railways, which depended for their efficiency upon

synchronisation, this kind of discrepancy created havoc. Imagine living in Cheltenham and every time you wanted to catch the 10.00 a.m. train to Paddington having to remember that it left at 9.52 a.m. Imagine trying to co-ordinate the timetables of different railway companies for a complicated cross-country journey involving connections.

Uniform time came about in Britain because of the railways. In 1840 the Great Western Railway introduced London (Greenwich) Time at all its stations and reprinted timetables accordingly. By 1847 all the major railway companies had followed suit.

Many towns objected to the coming of Railway Time, as Greenwich Time was first called. Places like Plymouth and Exeter, which were most affected because they were furthest west, held out against it for several years in angry displays of local chauvinism. The clock on Tom Tower in Oxford had two minute hands, one indicating Oxford time and the other indicating Greenwich Time. But it was a losing battle. Although uniform time did not become law until 1880, most public clocks kept Greenwich Time by 1855. Before the coming of the telegraph, the railway guard would have been responsible for carrying Greenwich Time on his pocket watch, ensuring not only the train's punctuality but conveying time's uniformity.

In 1852, Charles Shepherd's 'master' clock was installed at Greenwich. It was extremely accurate, and was corrected nightly against observations of the stars as they crossed the Greenwich meridian. Each second Shepherd's clock sent a signal to several 'slave' clocks at Greenwich which were synchronised with it, and every hour it sent a signal to a further 'slave' at London Bridge (even the language of clocks is imbued with colonialism). From London Bridge, the signal — which was no more than an electrical impulse — was sent all over the country via the telegraph wires. Bells rang, time balls dropped, guns fired simultaneously (give or take a nano-second) from Edinburgh to Bristol. This was the beginning of the Greenwich Time Service.

How did Greenwich come to be associated with time in the first place?

The quest for increasingly accurate clocks was not fuelled by some abstract desire for precision. By the late seventeenth century there was a pressing objective need for more accurate timekeepers in solving the problems of navigation which was still not a reliable art.

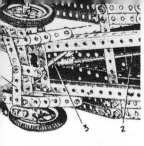

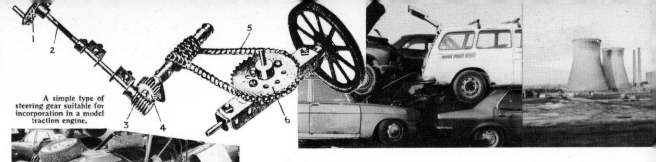

A simple type of steering gear suitable for incorporation in a model traction engine.

Originality is a feature of this simple but effective steering mechanism.

But who can afford a Range Rover? The Queen? The police?

We have to view the whole economy differently, and the argument becomes as much about ecology and energy as it does about more jobs and more creative work.

Today, if you throw away a car when it's ten years old, or done 80,000 miles, you're throwing away roughly the equivalent amount of energy that would be required to drive it another 80,000 miles. And, as we know, cars are designed to become obsolete in this way.

Cars are only one example. But we could design them in a way that was not only skilful and rewarding for the people who built them, but you'd also create a set of infrastructures in society where you could really maintain them.

Today, you can't take your car to a garage and ask them to repair your alternator, or replace a gear wheel in your gearbox. Instead, they'll take a major component out of the system and replace it with a new one, because these things are not designed to be repaired.

There's recently been a report for the E.E.C. ('On the Potential of Substituting Manpower for Energy'), which suggests that if cars were designed to last twenty years we'd not only conserve energy and materials but create 65 per cent more jobs. And these jobs would be all the interesting, diagnostic, fiddling type of repair jobs that human beings love doing, rather than experiencing the alienation of the shopfloor.

With my colleagues at Lucas Aerospace I have recently been involved in designing a hybrid power pack used for driving buses, trains and cars. We assembled it with bolted construction, and designed it so that it could really be repaired. With suitable maintenance, it would last twenty to thirty years.

You see, throwaway products also mean throwing away one of society's most precious assets: the skill, ingenuity, creativity and sheer interest of ordinary people.

The way human beings think, behave and act represents a degree of uncertainty. This uncertainty, which involves the ability to make quantum leaps with the imagination, is the essence of human creativity.

'I think I may make bold to say that there is neither any other mechanical or mathematical thing in the world that is more beautiful or curious in texture than this my watch or timekeeper for the longitude . . . and I heartily thank Almighty God that I have lived so long as in some measure to complete it.' John Harrison

51

To find your position at sea you need to know your latitude and longitude. Since ancient times navigators had fixed their latitude by measuring the angle of elevation of the sun above the horizon at noon, or the pole star at night. But finding longitude was not so easy. As late as 1700 there was still no precise method of ascertaining it.

In 1675 Christopher Wren chose the high ground and smoke-free atmosphere of Greenwich as a suitable site for the new Royal Observatory. The purpose of the Observatory was to make accurate star catalogues and tables of the moon's motions which were essential to the 'lunar distance' method then used for finding longitude at sea. This method was based on the assumption that the earth rotated on its axis at a constant speed, but this had not yet been proved. A secondary purpose of the Greenwich Observatory was establishing this proof and the experiments required more accurate clocks than had ever yet been made. The Great Clocks, as they were called, were installed in 1676 and established Greenwich as a source of accurate time measurement.

But the 'lunar distance' method of finding longitude at sea was complicated and prone to inaccuracy. In theory, everyone knew of a simpler and more reliable method. If you could be certain of the time at a standard reference point in the world, and you knew when it was noon where you were from the position of the sun, then a glance at a nautical almanac and a piece of simple arithmetic would give you your longitude. Because of the earth's rotation, the difference in time between the two places would amount to the difference in longitude. Greenwich, because of its location in the Port of London and its association with longitude and accurate timekeeping, was the obvious reference point for British shipping. If Greenwich is 0° longitude and your time is three hours later than Greenwich, then your position is 45° West. Easy.

But knowing the time in Greenwich was just the problem. For there was no clock accurate enough to keep Greenwich Time at sea, for months on end, thousands of miles from home.

The situation was pressing. In 1707, four Royal Navy ships ran into the Scilly Isles with the loss of 1,200 lives. The budding British Empire depended for its trade upon maritime power. But how could you hope to control the seas without accurate navigation?

In 1714, a petition went before parliament from 'several Captains of Her Majesty's Ships, London Merchants, and Commanders of Merchantmen', who proposed that a prize should be offered to anyone who could solve the longitude problem. The political nature of their need was clearly expressed:

'The discovery of longitude is of such consequence to Great Britain for the safety of the Navy, and merchant ships, as well as the Improvement of Trade, that for want, thereof, many ships have been retarded in their voyages, and many lost . . . the lasting honour of the British nation is at stake.'

Newton, who was a member of the Commons Committee that discussed the petition, went right to the point when he reported back to the House: 'For determining the longitude at sea, there have been several projects, true in theory, but difficult to execute: one is, by a watch to keep time exactly: but by reason of the motion of a ship, the variation in heat and cold, and the difference of gravity in different latitudes, such a watch hath not yet been made . . .'

The outcome was the establishment of a Board of Longitude, which offered a £20,000 prize for the solution of the problem. The watchmaker John Harrison finally received the full reward in 1765. He'd been trying for thirty years and won at the fourth attempt with his famous H4 Marine Timekeeper. It was accurate to within a second a day.

It's tempting to say that Harrison was only a whisker away from pinning down Newton's Absolute Time. On reflection though, the more precisely a watch 'kept time', the more time itself was defined by the characteristics of the watch. At first it had been the more or less regular forward movement of hours. Then came the minute hand. Now accuracy to within seconds. In this evolution watches and clocks increasingly refined the object of measurement they created, giving the illusion that Absolute Time was a measurable objective reality.

The British East India Company, who had assisted Harrison financially, were the first to insist that all their ships carry chronometers. The Royal Navy soon followed suit.

Before leaving its home port, a ship's chronometer had to be set against the standard time — Greenwich Time in the case of British ships. But since it was unwise to take the chronometer physically to the Observatory for fear of disturbing it, somebody usually took a good pocket watch. Time signals came about to avoid the tedious business of going to the Observatory. Initially these were dropped flags, fired cannons, searchlights or rockets. But from 1833 a time ball was dropped from the East Turret of Greenwich Observatory, every day at 1.00 p.m. Mean Solar Time:

'By observing the first instant of its downward movement, all vessels in the adjacent reaches of the river, as well as in most of the docks, will thereby have an opportunity of regulating and rating their chronometers . . .'

Eighteenth-century schooner used in Britain's coast trade

SHIPS, structures with which men make voyages at sea; which have increased in bulk from the open galleys of the ancients of 50 or 60 tons, to a timber ship of 5000. A first-rate Man-of-War being from 2500 to 2700 tons, and East India Ships from 1500 to 2000: the Merchants' shipping of Great Britain being 24,000 in number, carrying 2,600,000 tons. A first-rate carries 120 guns, 24 and 32 pounders, with a crew of 900 men; the length of her gun deck being 205 feet, and breadth 53 feet, the main-yard 106 feet, main-mast 124 feet, fore-mast 112 feet, and mizen-mast 112 feet.

The Rev. S. Barrow's *Popular Dictionary of Facts and Knowledge*, 1827

East Indiamen were better armed than most ships

Sloop of War, a vessel with two masts, carrying ten or twelve guns, and fifty or sixty men

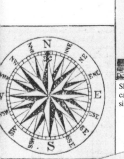
Sovereign of the Seas. She carried top gallant sails.

These frigates were built for the Far East run

In 1884 an international conference in Washington formally adopted Greenwich as the zero meridian for the world. The choice was a reflection of Britain's imperial power in the world at that time — three-quarters of the world's merchant shipping was already using Greenwich as the zero meridian. The inherent connections between finding longitude, the Greenwich meridian, and Greenwich Time, also led to Greenwich Mean Time becoming Universal Time for the world at the same conference. The political significance of this symbolic control over time was not lost on the French who persisted, in law, in referring to G.M.T. as 'Paris Mean Time retarded by nine minutes twenty-one seconds' until as late as 1978. Perhaps their intransigence paid off: although today's Universal Time is based on the mean of atomic clocks from twenty-four countries, it is none the less co-ordinated from Paris.

My hunch was right. Against British Rail's advice, I'm leaning out of a door window and I can't see the familiar black exhaust coming from the rear power car. There's no turbo charger whistle or heavy diesel shudder either. It's dead. Meanwhile the power car at the front has twice its usual workload, which means we're limping through the Yorkshire coalfield, barely capable of acceleration. The landscape in these parts is pretty dead as well. This is the thirtieth week of the miners' strike. Nothing is moving at the pitheads we pass. Winding gear is still. Rows of Coal Board trucks lie abandoned. The only sign of life is the occasional miner, fishing the strike away in bleak stretches of water hemmed in by track and slag-heap.

TAPERED SHANK

000 00 0 1 2 3 4 5 6 7 8 9 10

Now the main watchwords of Western science and technology are predictability, repeatability and mathematical quantifiability. A 'good' systems design in scientific terms is one in which you have eliminated uncertainty; whereas one which leaves space for human judgment, with its intuition and imagination, is perceived as defective because it leaves uncertainty in the system.

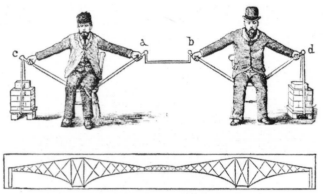

Living Model of the Cantilever Principle.

In my view, instead of seeing technology as something 'scientific', 'mathematical' and therefore 'neutral', we should see it as reflecting a set of values. Science and technology are cultural artefacts. They reflect the economic base of the society that gave rise to them.

Western capitalism is concerned with profit maximisation, which has led it to seek a means of gaining control over nature and eliminating uncertainty.

You can see this in the language we use. The very terms 'control over nature', 'manipulation of data', 'exploitation of natural resources' are all a reflection of the way we think about and view these things.

So, far from Western science and technology being neutral, it seems to me they reflect the predominant value systems.

What are these values that you're talking about?

Big Ben seems to symbolise the political power of British time even more than Greenwich Mean Time. The Illustrated London News artist, Smyth, saw this in 1858 when he drew the great bell with a Union Jack on top of it, on its way from the foundry to the waiting clock tower. Sixteen horses pulled his embodiment of British industrial time before vast crowds.

Big Ben conflates power over time and political power in an image that is now part of the psychological make-up of British national identity. Big Ben is different from the impersonal 'third stroke' or fifth pip. That familiar clockface (Ben is in fact the bell) asserts time with such analogical certainty that it seems to have a personality. As architectural guardian to the mother of parliaments, there's a way in which HIS pulse is both reassuring and democratic. The British sense of heritage and freedom seems bound up in the psychic resonance of his chimes and upright stance.

This kind old clock even authenticates news: a fanfare of chimes for the six o'clock news or an image to underwrite News at Ten. Memories of newsreels come flooding back in which 'London Calling' the world always seemed to be Big Ben calling. Standing there like a beacon, he seemed to be the sound and symbol of British Imperial Right for those away at war, for permanent exiles, and for our 'subjects' in the colonies.

Perhaps the benevolent symbolic power of Big Ben served to mask the harsh realities of British Time. For the export of industrial technology that characterised the expansion of British capitalism in the nineteenth century also involved the export of a discipline of time. In this process the very beat of British industrial life was arrogantly imposed on other cultures, with no heed paid to their indigenous times and rhythms.

ARRIVAL OF THE NEW BELL "VICTORIA," AT THE CLOCK TOWER, NEW PALACE OF WESTMINST

THE SOUDAN ADVANCE.

THE MARCH OF CIVILISATION: THE TELEGRAPH AND TELEPHONE ACROSS THE NUBIAN DESERT.

THE ILLUSTRATED LONDON NEWS, SEPT. 18, 1897.

THE SOUDAN ADVANCE.

The military expedition up the Nile, under command of General Sir Herbert Kitchener, is making steady progress. Garrisons have been placed at Dongola, Debbeh, Korti, Merawi, Abu Hamed, and Berber, with gun-boats patrolling along the river, and the route across the Bayuda Desert, to the south-east, is guarded by friendly native tribes. Berber has been occupied by General Hunter with a garrison of regular troops. At Ed Damer General Hunter has dispersed a hostile band led by Zeki Osman, and captured four large boats laden with grain. The construction of the direct railway line from Korosko, below Wady Halfa, due southward across the Nubian Desert to Abu Hamed, is rapidly continued, a hundred and sixty-five miles of it being already laid down. The Dervishes have mustered in strength at Metemmeh to oppose a final advance of the Egyptian army towards Khartoum. But it is not yet certain whether such an advance may not be deferred.

How wonderfully, in that region at the present moment, the conflicting forces of extreme barbarism and the effective inventions of modern science and skill are brought into play against each other! Some philosophical reflections, from this point of view, might be suggested by the sketches of our Artist, Mr. F. Villiers, which represent the unsophisticated natives employed under the direction of Egyptian army officers in laying the iron or steel plates and rails, or in carrying forward and fixing the telephone and telegraph wires; for such marvellous and mysterious operations must seem to the simple Moslem peasant like works of dire enchantment wrought by the power of Eblis or Sheitan, or of the "Djins" described in Arabian fables, still accepted by Eastern imaginative belief.

Egypt itself, during the nineteenth century, as well as the Soudan just now, has experienced from the introduction of European scientific methods, as a consequence of modern warfare, a prodigious extension or alteration of the mental horizon, compared with the formerly accepted limitation of the range of ideas in the Oriental mind. It is almost a hundred years since General Bonaparte's French army invaded Egypt. In the picture by M. Clairin representing the staff officers of that army viewing the stupendous architectural remains of the Temple of Karnak, we find another instance of the contrasts in the Valley of the Nile between objects of a venerable mystical antiquity and those which are the most recent products of latter-day civilisation.

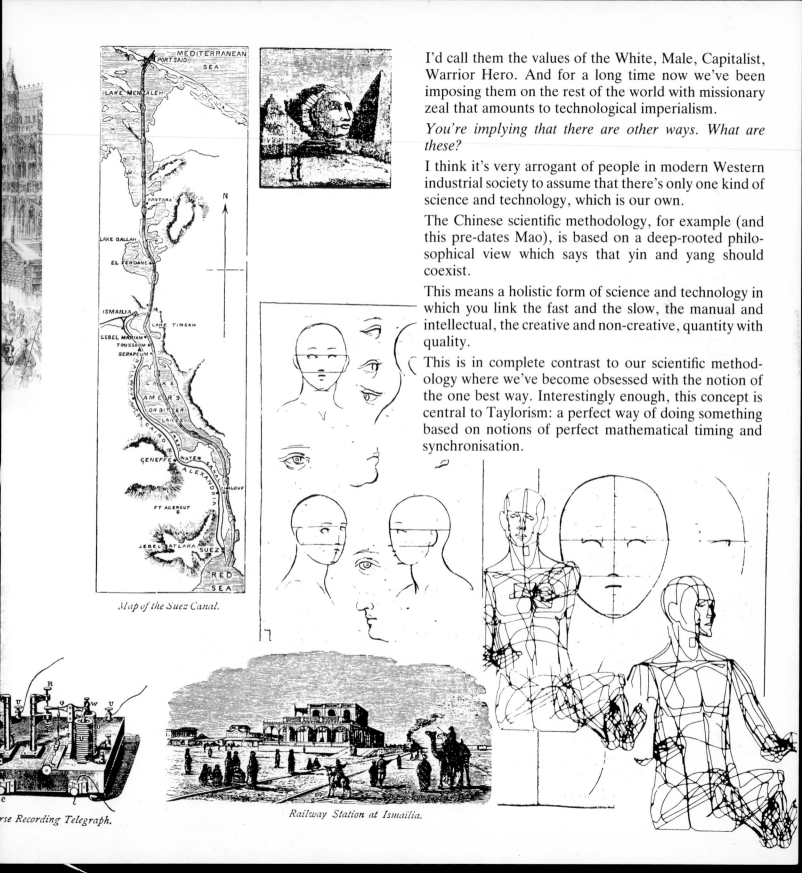

I'd call them the values of the White, Male, Capitalist, Warrior Hero. And for a long time now we've been imposing them on the rest of the world with missionary zeal that amounts to technological imperialism.

You're implying that there are other ways. What are these?

I think it's very arrogant of people in modern Western industrial society to assume that there's only one kind of science and technology, which is our own.

The Chinese scientific methodology, for example (and this pre-dates Mao), is based on a deep-rooted philosophical view which says that yin and yang should coexist.

This means a holistic form of science and technology in which you link the fast and the slow, the manual and intellectual, the creative and non-creative, quantity with quality.

This is in complete contrast to our scientific methodology where we've become obsessed with the notion of the one best way. Interestingly enough, this concept is central to Taylorism: a perfect way of doing something based on notions of perfect mathematical timing and synchronisation.

Map of the Suez Canal.

...rse Recording Telegraph.

Railway Station at Ismaïlia.

Today, with the Empire gone, the imperial Big Ben is a shadow of his former self. Lately he's to be found holed up in the B.B.C. World Service, where he chimes too much to a world he's lost his grasp on. Or worse, he's being sold in miniature quartz replicas to tourists in Parliament Square. Or just plain wrapped in his old age.

The decline of Britain as a world power is confirmed in the view from the train as we hobble into Leeds. Everywhere I look in this industrially scarred Yorkshire landscape I can see shut-down factories and disused mills. Stopped factory clocks hang in empty sheds. Paint peels from 'For Sale' and 'To Let' signs. The once audible pulse of industrial time has given way to an even worse option — the timelessness of the dole queue.

The train is at a standstill now, hovering at a red light outside the station while they let the Bradford train get out. The only thing that seems to be moving in this inner-city desolation is the glassed-in conveyor belt at Kay's Mail Order, which is delivering an endless stream of brown parcels from warehouse to dispatch. I can also see crowds of people milling around on the station platform waiting to board this clapped-out train for its return journey. And now my eye has been caught by the digital clock on top of the Evening Post building, silently sliding from 10.55 to 10.56. Forget about turn-round time, driver, you're ten minutes late already.

58

But you can't just import another approach to technology and time . . .

I think the alternatives are here within our own culture.

I once wrote an article called 'The Place of Daphne in the World of Apollo'. Now the Greek god Apollo was supposed to be the essence of rationality and eloquence, the God of Computing, you might say. Whereas Daphne represented the so-called female values of intuition and subjective judgment: values which have been eliminated from Western science and technology.

When workers at Lucas Aerospace were developing a corporate plan for socially useful production (as opposed to weapons systems), they made a major point of saying that they hoped that more women would come into science and technology. They didn't mean by this that they would come in as honorary males, imitation men. In my view, equality should never mean sameness. It should mean that the values which are said to be 'female' attributes are treated as equally important to the so-called 'male' ones, which now dominate.

If this happened, we'd end up with forms of science and technology which are far more humane and amenable to the human spirit, and would, in the long term, be more creative.

Yet another approach to technology could involve the reorganisation of time. In our present society there is a sharp division between work time and leisure time. Yet this idea that everything to do with work should take place during work time at the work place is completely at variance with the way we do things. I've found that when skilled workers develop new parts of machinery it's

often away from the work place. People have tremendous ideas when they're at home in the evening, simply musing. The basic concept of many great designs has come when people have been dreaming.

All the great scientists and engineers have spoken about this use of the imagination. Einstein believed the imagination to be far more important than knowledge. And Newton said on one occasion, 'I seem to have been only like a boy playing on the seashore, and diverting myself in now and then finding a smoother pebble or a prettier shell than ordinary, whilst the great ocean of truth lay all undiscovered before me.' Yet we're dominated by the idea that somehow masses of data in a computer will make us creative.

I think we should be able to work in the entire spectrum of space and time, both contemplatively and actively. I find too little space in modern industry for people to play on the seashore, and this is very important for human beings.

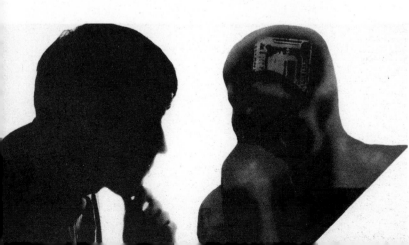

When all is hard, dark, sharp and black as crape and the wind keens a dismal threnody......

...and freezes each smile to a rictus of despair.....

.....at the Dead of Winter....then...here, entombed in John Innis potting compost No.1.....lies **HOPE**...

'Neath a polythene pall, dark days have been endured and the hard labouring of roots.....

And then, as blackbirds tune their mellow flutes and the bare buds curtsy to the wind.... and the throstle chimes his litany......

...trusty green blades pierce the sod....

...and a flourish of newly-minted trumpets gaze on a Springtime world, cleansed of care....

© Posy Simmonds 1981

HOLY DAYS

Since I have lived in cities, weather has come to mean what happens in strips of sky between buildings. But my childhood was spent on a farm in the country, where weather was everything. Each evening there would be a few minutes of reverent silence as we watched the shaman from the Meteorological Office perform his ritual magic on the television. How my parents would hang on his invocations: those forecasts laced with the magic words 'high', 'low', 'anti-cyclone' and 'depression' seemed to bring about instant gloom or elation. The weather map, with its jagged cold fronts and swirling isobars, became as familiar as bedroom wallpaper.

I was helplessly exposed to their anxiety. With an urgent tap of the barometer by the window, my father would strain to see high wisps of cloud, shaking his head as he muttered, 'Mares' tails', which promised more rain.

Will August storms flatten the wheat again? Will cutting drag on into September with the combine stuck axle deep in mud? How will we pay the grain-drying bill, let alone the wages? But it didn't rain on St Swithun's Day, and the rooks nested high in the trees this year. Perhaps that weatherman can make the sun shine.

For four million years human evolution has been influenced by the stark temporal parameters of the rotating, orbiting earth. Night and day, and the changing seasons, have imbued us physiologically and perhaps psychically with their rhythms.

An important condition of survival in this evolutionary process has been to live in harmony with the earth's seasonal moods. Knowledge of how and when these moods might change has been essential. In life for those who lived on the land, seasonal transitions weren't simply the times to stop harvesting and start ploughing, for example. They were the times to invoke the best, ritually, from the approaching season. They were the times for festivals, or holy days.

A rite of passage is a way of travelling from one condition of life to another without losing consciousness. We tend to think of them as ritual celebrations of the key transitions between the stages of a human lifetime, for the most part sadly lacking in our culture. For example, in some cultures young women are ritually initiated into their capacity to be a mother at the onset of menstruation, and in Japan there is a rite of passage to initiate men into middle age. But there are also annual rites of passage that initiate people through the thresholds of the year, from winter to summer, summer to winter, and so on. In this sense, customary festivals, like May Day, Hallowe'en or Guy Fawkes, are seasonal rites of passage.

Our calendar was once packed with local rites of passage that marked the key stages of the year. After Christmas and New Year came Plough Sunday, followed by Candlemas, Shrove Tuesday, All Fools' Day, Oak-apple Day, Hocktide, Harvest Home, Michaelmas and the 'Doleing Days' of November, to name but a few. In the round of seasonal activities on the land, the cyclical experience of the passing year was articulated by these festivals.

One of the effects of the Industrial Revolution has been a drastic thinning out of these festival rites of passage. City life has desensitised us to how intimately we are defined by the rhythms of the shifting seasons. The contrasts of the annual cycle have become bland: winter, at worst, is an inconvenience that looks to Dickensian London for its mythology. We tolerate higher fuel bills, the occasional burst pipe and the need to top up with antifreeze but unless we're snowed up or unemployed it's all somehow par for the course. Summer represents the holiday we hope we can afford: it's nicer when the sun shines in the city, but a bad summer is no disaster when you can buy strawberries all the year round. Our once cyclical experience of the year has been opened up into a shallow arc, barely distinguishable from a straight line, in which Shrove Tuesday becomes epitomized by a plastic lemon on a hoarding, Easter is D.I.Y. time, and May Day is a Communist plot.

In the face of surviving remnants of folk culture, townies become urban chauvinists. Customs are seen as quaint leftovers, preserved by country bumpkins whose rusticity

61

we capture on the Instamatic. Or we're just plain embarrassed by the self-conscious capering of local accountants dressed up as morris dancers.

These kinds of reactions aren't always wide of the mark. Often the meaning a festival might have had (the maypole, the sword-dance, mumming etc.) is eclipsed by an overbearing and misplaced sense of history. A dressed-up, cleaned-up, reconstructed past covers for an alienated present, with its stapled and stitched duty of doing things how they used to be done.

There are the grand exceptions. We still mark the Christmas/New Year transition with fervour. In recent years New Year's Day has been made an official holiday, sanctioning the day off people were taking anyway to recover from the night before. It's also becoming rarer to go to work during the odd days sandwiched between Christmas and New Year. The two events are increasingly becoming one long one, marking a return to the old 'Twelve Days of Christmas' – a midwinter holiday that celebrated the newly lengthening days at a time when little work could be done on the land.

Hallowe'en and Bonfire Night, the time of the onset of darkness and cold, is another transitional period that we still celebrate ritually. No amount of legislation has succeeded in extinguishing the need to make fire at this time of year, and the desire for ambivalent spookery keeps surfacing. But these events are far less socialised than they used to be. They tend to be atomised in suburban back gardens or officially controlled in faceless civic displays.

There are, of course, other 'days off'. But the generic term for them sums up precisely how far they've been removed from their original ritual function: 'bank holidays' are when money takes a rest. As rites of passage they've been reduced to exasperation in the tailback on the Exeter bypass.

This section is about the year's cycle and how we have marked and to a much lesser extent still do mark and celebrate its progress. The themes that underlie it touch on the sensitive area of biological determinism. They are based on the controversial inference that human beings have evolved a psychic need for seasonal rites of passage which modern industrial life largely denies them.

We start in May, with the Obby Oss festival at Padstow in Cornwall.

•

May Day at Padstow in Cornwall is widely recognised today as a unique and living folk event. Each year, the people of Padstow put on their whites, bring out their drums and accordions, deck themselves with bluebells and cowslips, and dance with the Obby Oss to the singing of the May Song.

The Obby Oss is a hobby horse, although he doesn't look much like one. In fact there are two Obby Osses – the red ribbon and the blue ribbon – followed by their respective parties, dressed in the appropriate colours.

The festivities start the night before, as midnight strikes, with the singing of the Night Song round the town. The next morning the Osses are brought out and by mutual consent each 'party' moves round the town without crossing the other's path.

There is a dancer inside each Oss, relieved at intervals, who prances to the drum beat and is teased, or baited, by a teaser with his club.

All day the singing and dancing goes on. After a number of verses of the May Song, a slower, sadder verse accompanies the Oss as he stops prancing and sinks down in death. Then, with a sudden beat of the drums, the Oss comes back to life, as the main song and the dancing resume. This rhythmic cycle of life, death and resurrection is repeated all day, all around the town, until the two Osses meet at the maypole in the evening in a ritualistic encounter of red and blue. Later, the Osses are put away to the sound of the Farewell Song. Summer has arrived.

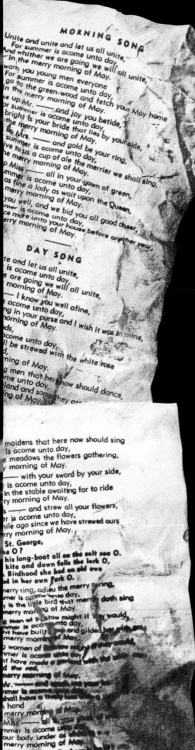

MORNING SONG

Unite and unite and let us all unite,
For summer is acome unto day,
And whither we are going we will all unite,
In the merry morning of May.

Warn, you young men everyone,
For summer is acome unto day,
Go to the green-wood and fetch your May home,
In the merry morning of May.

———and joy you betide,
For summer is acome unto day,
Bright is your bride that lies by your side,
In the merry morning of May.

———and gold be your ring,
Summer is acome unto day,
Give to us a cup of ale the merrier we shall sing,
In the merry morning of May.

———all in your gown of green,
Summer is acome unto day,
As fine a lady as wait upon the Queen,
In the merry morning of May.

———you well, and we bid you all good cheer,
———is acome unto day,
———more unto your house before another year,
——— merry morning of May.

DAY SONG

———te and let us all unite,
———is acome unto day,
———e are going we will all unite,
——— morning of May.

——— I know you well afine,
———is acome unto day,
———ng in your purse and I wish it was in mine,
——— morning of May.

———ds,
———come unto day,
———ll be strewed wim the white rose
——— ning of May.

———g men that here now should dance,
——— unto day,
———land and so ——— they are

———maidens that here now should sing
——— is acome unto day,
——— meadows the flowers gathering,
——— morning of May.

——— with your sword by your side,
——— is acome unto day,
——— in the stable awaiting for to ride
——— rry morning of May.

——— and strew all your flowers,
——— er is acome unto day,
———hile ago since we have strewed ours
——— rry morning of May.

——— St. George,
——— e O?
——— his long-boat all on the salt sea O.
——— kite and down falls the lark O,
——— Birdhood she had an old ewe
——— d in her own Park O.

——— merry ring, adieu the merry spring,
——— mer is acome unto day,
——— is the little bird that merrily doth sing
——— merry morning of May.

———a men of Padstow might if they would,
———mer is acome unto day,
———y have built a ship and gilded her with gold,
——— merry morning of May.

———g women of ——— might if they ———
———mer is acome unto day,
——— have made a garland with the ———
——— d the rest,
——— merry morning of May.

———Mr. ——— I ———
——— mmer is acome unto day,
———shall have a lively ———
——— hond
——— merry morning of May.

———Miss. ——— all ———
——— mmer is acome ———
——— our body under ———
——— merry morning of May.

——— St. George, O where ———

Peter Redgrove is both a well-known poet and novelist. Amongst his most recent books of poetry is *The Man Named East* (Routledge, 1985), while his most recent novel, a metaphysical thriller, *The Facilitators*, was published in 1982 (Routledge). His much acclaimed study of the human fertility cycle *The Wise Wound*, which he co-authored with Penelope Shuttle has been described as 'revolutionary', and he won the Prix Italia for Radio Drama in 1982 with his *Florent and the Tuxedo Millions*. He lives in Cornwall with Penelope Shuttle and their daughter Zoe.

PETER REDGROVE – MAY DAY AT PADSTOW

May Day at Padstow is a fertility rite that signals and arouses the energies of summer fertility both in the earth and in human beings.

It may seem foolish to recount this, but the first time I saw the Oss dying and coming to life again I was amazed to find that I was crying. Tears were spurting from my eyes like a child. And this was happening to the person I was with as well. I realised that what was going on at that festival was something that had never happened for me as a boy in church where they tell you it should. If I talk about it as a holy experience, everyone will dismiss it as religiose.

But if I talk about it as a *whole* experience, in which the four sides of myself . . . thought, feeling, sensation and intuition . . . were imaginatively participating . . . well, these are the most important experiences in people's lives.

Padstow is joyous. It's a celebration of the beginning of summer, and you suddenly feel those energies present. The shadow of the energies is there, of course, in the darkness of the Oss's mask, and in his death. But he rises again.

In times past we are told that the dancer inside the Oss used to carry a tar-brush. As he danced he would catch young women, dressed in white, and take them under the Oss to mark them with the tar-brush. This would be a sign of their fertility. Today the dancer occasionally takes women under the Oss, but I don't think he carries a tar-brush.

Rumours and accounts have it that in former times everyone would go up to the woods on May Day Eve to collect their cowslips and green boughs and they would make love together. And nobody ever knew who the fathers were to any children that resulted. I don't know whether any of this is true, but it's what I've been told happened.

How did the Padstow Festival originate?

Nobody knows. One of the things I noticed about the great mask and the music is that they look and sound African, and indeed there have been traders from Africa to this Cornish coast for millenniums past. The Phoenicians came here 4,000 years ago. Perhaps the form of the festival came from Africa, but who can tell?

Before the Civil War all England had its hobby horses and Cornwall was a thriving centre of this kind of thing. There used to be a great abbey church in Penryn, near Falmouth, called Glasney, and at the time of the Dissolution of the Monasteries it was savagely razed to the ground. The terms of the indictment spoke of 'pagan practices and Obby Osses'.

Implying that the Church would also have objected to Padstow?

Yes. They might well have called it Satanism or witchcraft. But Padstow survived all this. It's not a Victorian revival as some of these things are.

What was the essence of this experience for you?

Everyone has the kind of experience when trivial everyday life deepens and seems more solid and important. Sometimes this comes with ecstatic or religious intensity, and sometimes as mere contentment. And these are the most highly sought-after experiences. Half of

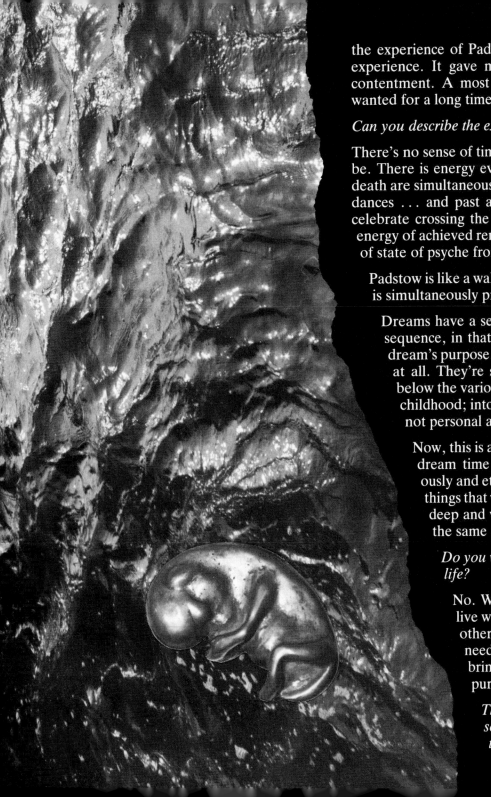

the experience of Padstow tunes us in to more complete levels of experience. It gave me a deep feeling of energy, harmony and contentment. A most extraordinary recognition of something I'd wanted for a long time which came on me quite unawares.

Can you describe the experience?

There's no sense of time or evolution there. Nor does there need to be. There is energy everywhere: the energy of opposites. Life and death are simultaneously present as the Oss dances, dies, resurrects, dances ... and past and future are simultaneously present as we celebrate crossing the threshold from spring to summer. I felt the energy of achieved renewal and it was familiar because it is the kind of state of psyche from which we began.

Padstow is like a waking dream, when everything that is necessary is simultaneously present.

Dreams have a sequence of exposition rather than a temporal sequence, in that the dreamer will alter time according to the dream's purpose. But the deepest kind of dreams have no time at all. They're simply time turned into space. And they go below the various levels of personal history, like yesterday or childhood; into a realm where you meet with imagery that is not personal at all, but universal.

Now, this is also the nature of ritual: one has come into the dream time where all important matters are simultaneously and eternally present. At Padstow all the important things that we need to know are present: the dream is very deep and very still because the ritual has been so much the same for so many years.

Do you value dream time more than the time of waking life?

No. We're creatures of both worlds, aren't we? To live wholly we have to give each world access to the other. The old legends in Cornwall say that the gods need us as much as we need the gods. As though we bring that other world into manifestation for some purpose.

This 'other world' that you are talking about: it sounds as though you see it as something more than my or your own personal unconscious.

Yes. Another word for it is the Implicate Order.

What is the Implicate Order?

I think that the experiences of timelessness, or the simultaneous present, that we have in dreams (or at festivals, in poetry, or art) are a close reflection of that reality which quantum mechanics is beginning to envisage for us as an objective reality.

The physicist David Bohm has suggested a picture of the reality of the universe in which every particle, energy and entity, is simultaneously implicated with every other one. There is a space/time continuum without past or future, where everything is present at once. Although this vision has not been accepted by all physicists, it has been the subjective experience through the ages of those we call mystics.

The simultaneity of this underlying structural reality, the Implicate Order, unfolds in our day to day experience into the 'common-sense' sequential time that we normally experience in the Explicate Order.

So you're suggesting that a festival like Padstow gives us access from this 'explicate' world to a greater reality beyond?

Yes. The Oss opens a door to that other world where all time is simultaneously present. It's a symbol reaching into the Implicate Order.

May 29th, 1949. Oak-apple Day in Great Wishford, Wiltshire. I'm wearing a toy policeman's helmet which is too big for me. The carnival procession is trudging along the village street in fancy dress – there are girls dressed as May princesses, the odd cavalier and a Mickey Mouse or two. Last year I was a fly-fisherman, but today I'm a law enforcer. They've tied a real policeman's truncheon to my belt, which is a great joke to them, but to a four-year-old it's a dead weight bouncing against my leg. As we traipse from the church to the marquee up behind the railway station, it jars my body, gradually nudging the rim of the helmet down over my eyes so that I can't see where I'm going. More to the point, I don't know why I'm going there.

It wasn't until I was older that I found out that oak-apples were those strange insect-ridden corky balls that hang from oak trees. But I still don't know why oak trees produce them as well as acorns. And why Oak-apple Day? Recently I read that it was to commemorate the Restoration of Charles II on May 29th, 1660. But there didn't seem to be anything patriotic about those rappings on our window at dawn with green boughs, and the all-night comings and goings to nearby Groveley Wood up on the hill.

Oak-apple Day in Great Wishford is a regional variation of the spring festivals that occur all over Britain in May.

Like Padstow, it's another way of marking the transition between spring and summer. But viewing these customary festivals simply as folkloric rites of passage rooted in a deep and fascinating pagan past, ignores how they evolved in response to immediate economic and social issues in the community.

For example, we tend to think that May Day festivals have only become politicised since the trade union movement took over May 1st as International Labour Day. The prevailing view is that the Red Flag co-opted what had been an innocent people's custom. Yet this assumption ignores the local class politics that have been at the heart of May Day for centuries.

Browsing in the family photograph album over the years, I've always thought that I was participating in just another village carnival. Nobody told me I was marching through Wishford to legitimise the right of local people to take dead wood for fuel from Groveley Wood. The political nature of the event is proclaimed in the slogan 'Unity is Strength' on the Wishford banner in a photograph taken in 1906 – a sentiment more usually associated with the organised working class than a quiet agricultural community.

In our conversation with Bob Bushaway, he laid as much emphasis on the immediate social and economic aspects of customary festivals, as on their origins as pagan rites of passage. This righting of the balance brings them down to earth and helps explain their continual evolution. As the title of his book, By Rite, *suggests, festivals are as much about rights as rites.*

67

BOB BUSHAWAY – MAY DAY

May Day and Bonfire Night are both rite of passage customs that mark important seasonal transitions in the year. Putting a maypole up involved taking a growing tree from the wood and bringing it to the village to celebrate the on-coming plenty of the summer. So May Day is almost the antithesis of Bonfire Night, which is about the coming of winter and the onset of hardship.

May Day used to be a period of great sexual licence. People would go off to the woods to collect their trees and green boughs, but once there they would enter into all sorts of temporary sexual liaisons which society did not normally accept.

Why isn't it like that now?

It was tamed and redirected. In the seventeenth century May Day came under severe attack from the Puritans who banned it by an Act of Parliament in 1644. In Philip Stubbs' 'Anatomy of Abuses', which was a Puritan tract against all kinds of merrymaking, there is a section called 'Against May' where he actually tries to measure the degree of sexual licence:

'Every parish, town and village, assemble themselves together, both men and women and children, old and young ... and go off, some to woods and groves, some to hills and mountains ... where they spend the night in pastimes. And in the morning they return, bringing with them birch boughs and branches of trees to deck their assemblies withal. I've heard it creditably reported by men of great gravity, credity and reputation that of 40, 3 score or 100 maids going to the wood overnight, there have scarcely the third part of them returned home again undefiled.'

The Puritans also objected to May Day, and other festivals, because of the way social hierarchy was set aside, so that all were commonly involved, from the highest to the lowest.

BOB BUSHAWAY is the author of *By Rite: Custom, Ceremony and Community in England, 1700–1880.* Awarded his Ph.D. while at Southampton University, he is now a University Administrator at Birmingham University. He is a regular T.V. and radio broadcaster, and is married with two children.

The Puritans found this offensive, much preferring strict gradations in society.

May Day did return, with the Restoration of Charles II in 1660, but it didn't have the same robust force. It had the old image, but the elements of sexual licence and social reversal went underground.

Then, in the nineteenth century, the Victorians overlaid a much more moral tone on the festival, emphasising its innocence. Instead of being a celebration of fertility it turned into a kind of commemoration of Merrie Englande. Girls taking part now wore white and held posies. John Ruskin played a big part in this.

What has this 'cleaning-up' done to the image of May Day today?

For the past sixty years folklorists have been rediscovering the pagan fertility tradition with its myths, rites and sexual licence. In my view, this has overshadowed the way in which May Day, and other customs, have been rooted in an economic way of life.

May garlands, for example, embodied the coming of summer. But they also legitimised the practice of knocking on doors around the parish and asking for money. At other times of year begging would have been an offence, but if it was done at Maytime with a garland, or collecting money for the guy, or wassailing at Christmas, it would have a powerful legitimation.

Also, the taking of the tree for the maypole highlighted the rights of people to take wood freely for fuel. This confirmed the extensive medieval rights to wood usage, including both the taking of growing timber for building and repairs, and dead wood for fuel. A classic 69

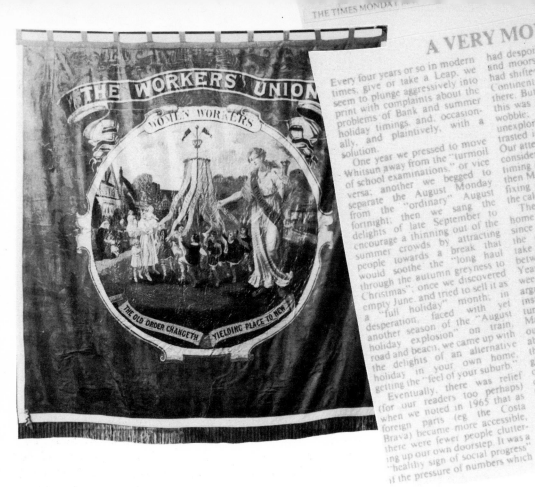

A VERY MOVABLE FEAST

Every four years or so in modern times, give or take a Leap, we seem to plunge aggressively into print with complaints about the problems of Bank and summer holiday timings, and, occasionally, and plaintively, with a solution.

One year we pressed to move Whitsun away from the "turmoil of school examinations," or vice versa; another we begged to separate the August Monday from the "ordinary" August fortnight; then we sang the delights of late September to encourage a thinning out of the summer crowds by attracting people towards a break that would soothe the "long haul through the autumn greyness to Christmas", once we discovered empty June, and tried to sell it as a "full holiday" month; in desperation, faced with yet another season of the "August holiday explosion" on train, road and beach, we came up with the delights of an alternative holiday in your own home, getting the "feel of your suburb".

Eventually, there was relief (for our readers too perhaps) when we noted in 1965 that as foreign parts (eg the Costa Brava) became more accessible, there were fewer people cluttering up our own doorstep. It was a "healthy sign of social progress" if the pressure of numbers which

had despoiled our downs, coves and moors in the high summer had shifted to other parts of the Continent, to do the same thing there. But we soon realized that this was actually an unpatriotic wobble; deserting one's own unexplored "marvellously contrasted island" was frowned on. Our attention shifted to a regular consideration of the role and timing of the Bank holiday, and then Mr Heath's late August date fixing really put the cat among the calendar pigeons.

The birds have been coming home to roost, braving the cat, since that decision, coupled with the developing inclination to take a clump of national holiday between Christmas and the New Year, a sort of winter wakes week. Finally, we got, in 1978, arguably the first politically instead of religiously (or agriculturally) motivated holiday in May Day. Others have taken on our aggressive, or plaintive, role about that date. They wish to see the celebration, "the most gloomy spot on the vacation calendar," moved to another date, like St George's Day, or the Queen's official birthday, or even the preferred current date of the English Tourist Board, which is sometime in June. Or they would like September (We have been there before.) Or almost any time other than May.

The national disinclination to do anything conveniently ordered is on a par with disinclination to show solidarity with the world's workers by taking to the streets on May Our calendar has already separated, like our religion the political mainstream world.

On this newspaper, we very much in favour, point, of well ordered for all, leisure, elbow the beaches, saints days ances, a fair day off the day's work, the Cost peace during school ations, patriotism, safe roads, peace on the tranquillity on the pre a happy June and autumn. We acknow not all of these con Until it can be so of May Day happens to first day of the year w enough sun for us warmly in the solidarity of spirit body, we will se movable, which quo. In any case it be celebrating wh saint's day, it is William Pitt (J memory of that worker, we shou Mayday....

Let's scrap this holiday

WHAT a washout. The May Day was not simply wet, it was un Only the snooker tournament on TV from total disaster.

The idea of the May Day strenuously promoted by Mr Micha was to make a public celebration of rejoicing.

It is a political occasion. As shows by putting on a propaganda b every year. As Sheffield's Left-wing demonstrated by flying the Red Flag Town Hall.

Why should the rest of us have with this Salute to Karl Marx? W have a Bank Holiday at the end o

We should end the May Day and establish a new holiday in it the middle of June. This would two magnificent achievements, v Napoleon at Waterloo and vict Falklands.

America has a number o commemorating national events. and New Zealand have Anzac Da the heroism of Gallipoli. Cana National Day. So has South Afri we.

example survives at Great Wishford in Wiltshire, where wood gathering rights are legitimated on Oak-apple Day. There were a number of major disputes in the eighteenth and nineteenth centuries in which landowners challenged this right, to which people would claim in their defence that it was Maytime and so legitimated.

Why did the Labour Movement choose May Day as International Labour Day?

It's more that May Day chose the Labour Movement. Unlike Easter, Whitsun or Christmas, May Day is the one festival of the year for which there is no significant church service. Because of this it's always been a strong secular festival, particularly among working people who in previous centuries would take the day off to celebrate it as a holiday, often clandestinely, without the support of their employer. It was a popular custom in the proper sense of the word, a people's day, so it was naturally identified with the Labour and Socialist Movements and by the twentieth century it was firmly rooted as part of the Socialist calendar.

It's only recently that the state has recognised May Day as a bank holiday, for the first time since it had royal support back in the Elizabethan court. And there's been a big battle over this: May Day was seized upon by the right as something foreign and left wing, but this entirely misses the continuity of its roots in our cultural tradition.

Calendars are a means of measuring the timing of the year for the community. They timetable the annual rites of passage (whether these be pagan festivals, Christian church services or political celebrations) to fit the seasonal and monthly movements of the earth and moon. And these ritual thresholds pace the intermediate stretches of everyday working life. In this way, the religious and secular dimensions of a calendar are complementary.

The basic natural clocks for calendar makers have always been the apparent movement of the sun and/or the moon. Nomadic and hunting cultures tended to base their calendars on the moon, because when you move by night moonlight is essential. Farming cultures, which evolved later, tended to base their calendars on the movement of the sun, which governed the time of planting and harvesting crops. In the next section, 'Moonshine', Reter Redgrove and Marie-Louise von Franz take up the psychological implications of the influence of sun and moon on culture and calendars.

The trouble for calendar makers has always been that the moon and the sun are not neat timekeepers. The movements of the earth round the sun, and the moon round the earth, are regular, but the arithmetic is a schoolboy's nightmare: a mélange of long division with mind-bending remainders.

In our Gregorian calendar we have twelve 'moonths' of uneven length which jostle each other to make sense of the mathematically awkward lunar cycle of twenty-nine days, twelve hours and forty-four minutes, while still adding up to the 365 days of the solar year. Then there are fifty-two times seven-day weeks, seemingly unrelated to the months, which add up to 364 days, one day less than the solar year, which is in fact not 365 days but 365.2422. This hodgepodge of competing rhythms reflects both the calendar's complex cultural evolution, and the fact that there are no clean sums to be done when it comes to measuring 'nature'.

Julius Caesar confronted these problems in his major reform of the calendar in 46 B.C. The Romans had inherited the Egyptian 'Wandering Calendar' which had 365 days. For many centuries this calendar had been gaining 0.2422 days per annum on the solar year. Every four years it had become a day further out of sync with the seasons, creating a cumulative havoc in administering the civil and religious year. The problem was solved by adding (intercalating) an extra day every fourth year, which we've come to know as the leap year.

But even this wasn't quite right because the calendar was now losing 0.0078 days per annum against the solar year. By the sixteenth century fixed-date festivals were becoming significantly out of phase with the solar year. In 1582, Pope Gregory cancelled ten days to accommodate the error. The complex nature of the arithmetic can be seen in the steps he took to ensure against future errors: he decreed that all centurial years (1700, 1800, 1900, etc.) except those divisible by four (1600, 2000, etc.) should be years of 365 days, thus losing the odd leap year and making no further adjustment necessary for another 3,000 years.

The English reaction to Gregory's reform underlines the ideological nature of calendars. In the climate of Reformation England, a calendar shaped by a Pope was unacceptable. It suggested a concession to Catholicism. It

Simmonds 1983

wasn't until 1752 that the English adopted the Gregorian calendar. Eleven days were dropped in September of that year, and there were riots because people thought that they were losing eleven days of their lives.

In the broadest sense, power in society means power over people's time. When governments legislate the length of working days or sanction slavery, they are fundamentally shaping the social experience of time.

Likewise, control over the calendar is a form of power over social time – the temporal rhythm of the year – and almost invariably lies in the hands of religious or secular ruling groups – priesthoods or governments. In fact, the temporal structure of the year is so integral to the structure of a society that new political orders have often been accompanied by major calendrical reforms. The resulting calendars almost always bear the ideological imprint of their makers as if symbolising their power over society's time. In the Julian calendar, the power of the reformer and his successor were immortalised in the renaming of two months in their honour – July and August.

In the case of the French Revolutionary calendar, the reform went beyond symbolism to a fundamental reorganisation of social time. 'Reason' was the watchword of this revolution and it imbued their attempted decimalisation of the calendar. Any association with Christianity and the Catholic church, which had shaped the calendar since the Julian reform, was thrown out. Three ten-day weeks formed each of the twelve new uniform thirty-day months, the five spare days being grouped together at the end of the year, with a sixth one added on leap years. Moreover, each day was divided into ten hours, each with one hundred decimal minutes, which in turn comprised one hundred decimal seconds. Christian year dating was abolished and time was 'restarted' from the first day of the Republican Era – September 22nd, 1792 (the new New Year's Day). Days were renamed Primidi (first day), Duodi (second day), etc; months were renamed after the characteristics of the seasons – Messidor (harvest), Thermidor (heat), Germinal (seeds), etc.

The French Revolutionary calendar lasted until 1806, when Napoleon reverted to the Gregorian calendar. His reasons were probably as much practical as they were a conscious political distancing from the values of the French Revolution. Calendrical conventions are so deeply rooted in the ritual traditions and working rhythms of societies that they resist wholesale political interference. Stalin faced the same problems when he tried unsuccessfully to banish Sunday and instigate a six-day week. On a smaller scale, the Puritans banned May Day and Christmas in England during the seventeenth century, but these rites of passage asserted themselves and re-emerged with the Restoration.

Yet while the structure of our calendar is the result of an unbroken evolution from Ancient Egyptian times, based on astronomical realities, the Industrial Revolution has had a radical levelling effect on its local profile. We still have the same framework of months and weeks, but customary festivals don't articulate the cyclical movement of the year as they once did. The move from the country to the city in the past two centuries has meant that changing seasons no longer have such a direct effect on people's lives. While the psychic need for seasonal rites of passage perhaps remains, we've been severed from their immediate economic necessity and ritual purpose.

BOB BUSHAWAY – THE INDUSTRIAL REVOLUTION AND HOLIDAYS

What impact did the Industrial Revolution have on customary events?

Before the Industrial Revolution the local customary calendar would have specified your days off. Each parish had its own feasts and wakes based on its particular saint's day and other events in its church calendar.

This meant that each area had its own haphazard holiday pattern. If you lived in a good area you could go from one parish feast to another in succession and this was found to be very disruptive to the work discipline required by the new factory-based economy. Regularisation of holidays was the result: employers much preferred their workforces to take all their time off in a single week rather than put up with lots of sporadic holidays that bore no relation to the work pattern.

This is very much how Wakes Weeks developed. And it's still true in my parish of West Heath where the large B.L. plant at Longbridge shuts down for the last two weeks of July. This affects the whole local economy: go to a butcher's at this time and he won't have the cut you want because he's only keeping a limited stock when the Longbridge workforce aren't there.

The advent of the railway also undermined local calendar customs. Cheap fares and excursions meant people could have a day off away from their accustomed environment. And there was a great appeal in going further afield. The mass exodus to places like Blackpool and Brighton began at this time. It was the start of the Great British Holiday. People were no longer content to see their time off as a localised event now they could broaden their experience by leaving their parish.

So the time of year you take your holidays today is possibly less significant than formerly, when it was bound up with a festival as a rite of passage on a date specified in the local customary calendar.

Today we take paid holidays for granted but, like the eight-hour day, they were hard won. Even as late as the 1950s only a minute percentage of the workforce enjoyed these benefits. 73

'The study of the origin of words may be regarded as a sort of archaeology of our thought process, in the sense that traces of the earlier forms of thought can be found by observations made in this field. As in the study of human society, clues coming from archaeological inquiries can often help us to understand the present situation better.'

David Bohm: Wholeness and the Implicate Order

HOLIDAY – Day of cessation from work, or of recreation; period of this, vacation. (From Old English: haligdaeg. See HOLY and DAY.)

HOLY – Consecrated, sacred. (From Old English: halig, meaning whole.)

WHOLE – In good health, organic unity, complete system.

HEALTH – (From Old English: haelth (whole-th).)
Oxford English Dictionary

We're going on holiday tomorrow. We'll leave the farm and the anxieties of the harvest behind for my father to worry about. We've hired a caravan by the sea, where for three weeks I'll scour the sand dunes for empty Corona bottles and spend the proceeds in the corrugated iron shed marked 'Ices and Amusements'. And off we'll go, into another of those endless wet summer days in the back of the car, paced by the tired to and fro of windscreen wipers and the sound of raindrops on the canvas hood.

•

Most people speak of their holidays as a time 'to get away from it all', 'to relax and let go', 'to forget about everything'. Central to this expectation is the concept of leisure time in sharp contrast to work time, when the ravages of work are healed and recoveries are made.

The 'holi-' of our modern holidays still means wholeness and health, but it refers more to the physical and emotional health of the individual than to the spiritual health of the community that customary holy days ensured.

At a festival like Padstow, the 'sacred time' of May Day, while ritually distinct from the 'profane time' of the rest of the workaday year, is none the less inseparable from it; the purpose of it as a rite of passage is to bid farewell to the spring and initiate the summer. Here, the 'whole' in holy day means 'organic unity ... complete system', and embraces our overall experience of time.

The Industrial Revolution can be seen to have encroached on and drastically thinned these kinds of holy days (customary festivals) and replaced them with holidays.

The effect of the modern holiday, while broadening our horizons and giving us a good time, is to break up the flow of the year into work with its weekends and bank holidays and the annual package of getting away from it all.

In our conversation with Edmund Swinglehurst we talked about Thomas Cook and the rise of the Great British Holiday in the nineteenth century.

Edmund Swinglehurst trained as a painter under Léger – since then he has been schoolmaster, insurance clerk, photo processor and advertising executive. Today he is a writer. His books include *The Romantic Journey – The Story of Thomas Cook and Victorian Travel* – an outcome of his long association with Thomas Cook & Son.

EDMUND SWINGLEHURST – THOMAS COOK AND THE RISE OF THE GREAT BRITISH HOLIDAY

I think there is a seasonal rhythm in the personality and psyche of human beings that needs to be expressed at certain times of the year. These were the original holidays which coincided with the times of important changes in nature, like the solstices of midsummer or Christmas, or the equinoxes of spring and autumn. They involved a spontaneous need to celebrate and take time off, and were quite short (two or three days) but also quite frequent.

The annual holiday of a week or two weeks in one lump is a development that took place in the nineteenth century. It didn't really become a reality for most people until the 1920s. Annual holidays for working people were not the gift of beneficent employers but the result of a long hard struggle.

Where does Thomas Cook come into this story?

Thomas Cook was born early in the nineteenth century in Derbyshire. He was a working-class boy and had no social advantages in life. He started work as a gardener's boy at the age of ten. He had no formal schooling, although he received his education from local lay preachers, which may have given him his early sense of responsibility towards society. In his late teens he became a Baptist preacher and travelled extensively, often up to 2,000 miles a year on foot. On these travels he soon came to see at first hand the terrible conditions of poverty that most people lived in. He was also struck by how they tried to seek oblivion in drink and his religious convictions inclined him to do something about it. So he became a member of the Temperance Society.

Thomas Cook holidays started here: he thought that by giving people the opportunity to fill their time with travel and see places they'd never seen before, he might take their minds off drink. His first tour was in 1847 when he took a Temperance party on the train from Leicester for a picnic excursion in Loughborough.

Was there much pleasure to these excursions with so much emphasis on moral improvement?

In fact, most of his tours had people who drank on them. For example, Lord Shrewsbury, the owner of Alton Towers, opened his estate so that Thomas Cook tourists could admire the magnificent house and grounds. Later he wrote to Cook saying he couldn't stand any more of it since half of the party had been drunk.

Cook's large-scale tours started with the Great Exhibition of 1851. Working from his offices in Leicester, he planned to bring thousands of workers down to London from the grimy factory atmospheres of places like Middlesbrough and Sheffield to see the Exhibition. The idea was to inspire them with all the wondrous machinery and inventions, and at the same time create in them a political awareness that *they* had produced the wonders they were admiring, and consequently deserved better conditions.

75

He even started a newspaper called the *Excursionist and Exhibition Advertiser* in which he wrote articles asking why only the sons of the rich should have the freedom to travel and enjoy themselves. He was unpopular with the establishment of the day. The brewers didn't like him because he encouraged people not to drink. And at first employers didn't like him for encouraging workers to take time off – on occasions he'd actually arrange for trains to be at the factory gates and, as the workers came out, there he'd be with a brass band urging them to come to the Exhibition.

Cook ended up in bringing 140,000 people to London to see the Great Exhibition, which he couldn't possibly have done without the eventual co-operation of the employers. If he'd antagonised the railway barons, for example, he wouldn't have been able to move people around. So he wrote a number of articles to persuade employers that if their workers were allowed to visit the Exhibition so as to see the results of their labour they'd be inspired to work harder and feel a greater sense of participation in the country. This tactic succeeded, even to the extent that many employers actually paid for their employees to go.

So a trip to the Great Exhibition was seen more as an education than a holiday?

This touches on the heart of Victorian morality. When the Industrial Revolution got under way it was seen as a God-inspired event that was making Britain the world's leading country, and it was felt that it was the duty of everyone to work. Work was ennobling and any leisure, especially working-class leisure, had to be justified. Gone were many of the disruptive and all too frequent days off for festivals. Now holiday time had to have a purpose. The rationale was that leisure time, itself a new concept, should improve either your health, your mind, or both.

So leisure was seen as a kind of work?

Yes, but the traditional subconscious needs for a holiday as a time for fun and enjoyment are irrepressible. The development of the day trip to the seaside shows this clearly.

For centuries the seaside had been largely ignored as a place where only fishermen lived and storms happened. When the middle classes started going to the coast early in the nineteenth century, the resorts began to develop. But the moral issue of leisure time was such that they had to justify going there. Going to a spa town like Harrogate or Tunbridge Wells to take the mineral waters was acceptable on the grounds of health. Next it became acceptable for them to go to the seaside for their health. Many books were written about this: Drs Russell and Granville described how sea water was a cure for almost everything from infertility to athlete's foot. Not to swim in, but to drink. Five pints a day were recommended, and if you didn't like the taste of it you could have it with milk or brandy.

Now, in the 1850s and 60s Cook began running tours all over the country, and the working-class day trip to the seaside began to flourish. But now the rationale that they were taking these holidays for health reasons was dropped. Workers went to get away from their working environment. If you lived in a village or a tightknit urban community, you had to toe the line and behave in accordance with Victorian morality. But in a place like Brighton or Blackpool you could find anonymity and freedom from moral constraint. Nobody knew who you were and you could just mingle with others along the promenade. Indeed many

people didn't get as far as the promenade. They used to hang around in the pubs near the railway station, kissing, cuddling and cavorting in a way they couldn't do publicly at home. This sort of behaviour, although shocking to the middle classes, broke down the feeling that a holiday had to be justified.

How did the shocked middle classes react to this invasion of their holiday resorts?

They went abroad, following in the footsteps of their superiors who used to take the Grand Tour in the eighteenth century. And Cook catered for this as well.

He took them to places like Switzerland and Italy where the aristocracy had gone. Switzerland, because its soaring mountains, waterfalls and forests satisfied the romantic yearning of the age; and Italy, especially Rome, because they saw the Roman Empire as the predecessor of the British Empire – they thought of themselves as modern Romans.

But now education replaced health as the main purpose of the holiday. You see, these middle-class people had social aspirations, and felt that the way to get on in society was to be seen to be like people superior to them. Since the aristocracy were thought to be well-educated, the middle classes emulated them, beginning a great vogue for learning about places abroad. With the help of the Baedeker guidebook, they learned about churches, castles and monuments of all descriptions. It became very important to remember exactly which places they'd been to, how many stained-glass windows there were in Chartres cathedral and so on – it was an epoch of acquisition.

If you were a person getting on in society, you'd possess a nice home with nice furniture, a piano, a few statues, etc. In the same way, a holiday became part of your possessions. When you went abroad you'd collect both experiences and objects, and there was status in what you collected. If you saw the Colosseum by moonlight you'd be one-up on someone who'd only seen it by day. And you'd be sure to buy souvenirs – alabaster models of marble statues and so on.

Did the middle-class people who took these holidays abroad experience the same release from everyday morality and convention that the working class experienced at the seaside?

Yes. But not so explicitly. One of the reasons for Cook's success was that he always tried to show that he was very respectable and that nothing untoward would happen. Middle-class Victorian life was very constrained: Cook offered a freedom that involved climbing mountains, walking freely round the ruins of Pompeii and, if you were a woman, climbing over glaciers in your crinolines. It was a kind of physical freedom that people didn't have at home which made Cook's tours especially popular with women.

But not the sexual freedom we have come to associate with the modern holiday?

I'm sure it happened, but not as visibly as either today, or on Blackpool sands in those days. The romantic stories in the magazines of the time suggest something of the sort – secret meetings on mountain tops or at lakesides, breath coming more rapidly as hands were held in mounting excitement. Later in the nineteenth century Cook used to organise cycling tours. In the advertising he said that these tours were perfectly respectable for ladies, and if they were on their own they might find an agreeable companion! Perhaps these were the first 'singles' holidays.

77

When did the working classes start going abroad for their holidays?

Not until after the Second World War in any significant numbers. By the 1950s people finally became legally entitled to a holiday with pay, so suddenly a market of many millions of people with money in their pockets developed. A host of tour operators sprang up in response, to sell the idea of foreign holidays.

And did this new wave of tourists go guidebook in hand, as the middle classes had done fifty years earlier?

Initially, yes. The educational habit persisted because the travel agents organised excursions to all the famous places which people felt that they ought to see. And the souvenir business persisted as well. But as the years went by the improving aspect of foreign travel became less important. The brochures of the 1960s show this. They present the holiday abroad as a new and thrilling experience which is not necessarily about collecting solid objects and getting an education. They play up the romantic element, enticing the holidaymaker to come to a wonderful new land, where they'll meet all sorts of new friends and create new relationships. Instead of just coming back laden with photographs and souvenirs, they'll return with enough experience tucked away in their minds to see them through the rest of the year.

The essence of a holiday is not learning about other cultures, although many people like to do this. It's to break free from the rhythms of work time and the constraints of everyday space. On holiday we enter a different time and space where there is, hopefully, no schedule and the limitless space of the sea. The modern holiday is a time out of time, like the day trip to Blackpool, but long enough to forget work time.

Do you think the experience of a modern holiday is different in kind from the experience of seasonal festivals?

The difference is that the modern holiday lacks such a precise purpose. It's the same in that it offers a different time and space, but it doesn't satisfy the need in us to express seasonal changes. And however sophisticated and industrialised life becomes, you can't change this need. Since the Industrial Revolution we've been going through a period when we disregard these rhythms, but I think we may well grow out of this.

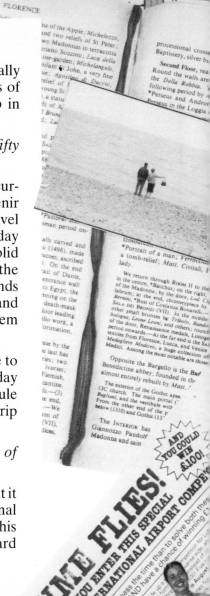

But in late August the dew starts getting heavy, so we have to wait until noon, hoping for a good breeze which will start the ears off in their crackling chatter, telling us they're dry enough to cut.

Then it's all through the afternoon and into the evening, hoping the combine will pick up the rain-flattened crop. Round and round we go, in the dust and the sweat, with the clatter of the cutter bar and the deep whine of the threshing drum, filling sack after sack, through to tea-time with Thermos flasks and cake on the stubble. Then off again, in ever decreasing circles, anxiously watching the sky, waiting as dusk falls for the screech of the clutch as the dewy straw bird's nests round the drum, so tight that even my sheath knife won't cut it.

I don't recall that we ever celebrated getting the harvest in. Perhaps, by the 1950s, there were just too few of us to warrant any festivity.

Once there would have been forty of us in this field – men, women and children – cutting, tying and stooking the corn by hand. This year I'm big enough to help, which makes just me and the driver up on top of this old Massey Harris combine, about to start on ten acres of wheat.

The trouble with combines is that you can't cut when the wheat is damp. The moist straw will bind round the threshing drum and bring the machine to a halt. And if you're lucky enough to avoid this, the grain itself will almost certainly be too wet to store. There wouldn't have been this problem in the days of reaping and binding: the wheat would have been cut much earlier, and stooked in sheaves to ripen and dry in its own time.

MARIAN GREEN is the Editor of *Quest* magazine and is a full-time writer and lecturer specialising in folklore and magic. She graduated from London University with a B.A. Honours Degree in Humanities and now is involved in teaching various modern aspects of magic to students from all over the world. She has made a number of radio and T.V. appearances in programmes associated with witchcraft, E.S.P. and the paranormal throughout her career.

MARIAN GREEN – HARVEST

One of the reasons that school summer holidays are so long is that before the days of the combine harvester, children were needed to help with the harvest. They might start as early as June, helping with the hay harvest. Then they might spend a few weeks picking stones from the fields until it was time for the corn harvest, when their main job was stooking.

As soon as the first corn was in and dry, bread would be made from it. The celebration of the first loaf made from the new grain was Lammas (meaning loaf-mass in Old English). It was held at the beginning of August.

Today we know Wakes Weeks as the traditional time for the annual holiday in the industrial north of England: the time, varying from town to town, of the mass exoduses to the seaside. We tend to forget that Wakes Weeks used to be the time when people took a week or two off work to help with the harvest, and that the Wake was the festival celebrating the dead spirit of the newly cut corn.

The old pagans believed that there was a King of Summer and a King of Winter, and that they fought each other at midsummer and midwinter. The King of Winter wins at midsummer; he carries the reaping hook with him which cuts the corn and is the symbol of death.

The King of Summer is the Harvest King, the King of the Corn. He is the symbol of the life spirit that is told about in the song 'John Barleycorn', and he is cut down, sacrificed, so that the people can live.

At the end of harvest, the last sheaf of corn was taken and woven into the shape of a man or woman. This was called a corn dolly and was supposed to keep the spirit of the King of the Corn alive through winter. The grain of the corn dolly was then sown the following spring in amongst the rest of the seed corn.

The Wake was the festival of the 'dead' spirit of the corn, the celebration of the end of harvest, which culminated in a great feast known as the Harvest Home.

Today there are few vestiges of the original custom. In some areas the Harvest Supper survives. And we see traces of the corn dolly in the sheaf on the church altar at Harvest Thanksgiving.

MODERN FARMERS RELY ON

THE END OF THE HARVEST: THE LAST SHEAF. A CORNISH CUSTOM.

Shell Chemicals

SHELL NO.2
COMPOUND FERTILISER
21-14-14

FERTILISERS IN LIQUID FO

The
pattle
is on
again

AND THIS ONE'S EXTRA!

in straight nitrogen fertilizers

s, never bogs down, and
werhouse will carry a load
an be fitted with a

st broad-leaved
has started.

scentless mayweed, knotgrass and cleavers
And because it has been proved to kill
ly than any other weed
strongest

The wheatsheaf on the altar at the Harvest Festival was the golden centrepiece, and it came from our farm. There it stood amidst its supporting cast of chrysanthemums, leeks, cauliflowers and baskets of brown eggs lovingly arranged, the glowing focus of our ardent singing of 'We plough the fields and scatter'.

All this thanksgiving tended to obscure the bleak reality it referred to. Some years the weather was so bad that we had to plough whole fields of uncut wheat back into the earth. And one year our sitting-room was stacked wall to wall with open sacks of grain, in a futile attempt to dry it out before mould set in.

But at least this praise bore some relation to a harvest that might have been seen or heard going on near by. Today, in the cities, the churches still have their Harvest Thanks-givings, but the moment has lost its significance. The freezer compartments in supermarkets provide us with an endless harvest all the year round, while the produce on the altar is more likely to consist of plastic flowers and a few packets of digestive biscuits, than a real wheatsheaf.

This same 'plasticification' now characterises Hallow-e'en. Yet instead of trivialising this festival, the increasing popularity of joke shops selling rubber webs, model witches on broomsticks and plastic pumpkins perhaps signals a revival.

What's the significance of Hallowe'en?

It's a kind of Harvest Festival of the spirit. The grain, fruit and vegetable harvests are safely in. Now you can gather in the harvest of your ideas and achievements of the past year, seeing which ones you want to keep and which to throw away.

It's a time when you gather your friends and family around you, and when the ghosts of your family come and join you.

Hallowe'en marks the end of the old Celtic year and the beginning of the new. The ancient Celts celebrated it as a festival of the dead. The Christian Church took over the same idea with All Souls' Eve. The spirits of the dark side of the year were thought to roam at this time, but it wasn't a festival of evil or harm. Dead people were thought to be wiser, so these spirits could impart knowledge and wisdom.

Often a feast would be held with an empty place laid for the family's most honoured guest . . . This might be an ancestor or even a god or goddess (our 'original parents'). They were thought of as guests and as the evening wore on and people got a bit drunk they might feel that they could see them as ghosts.

Nowadays Hallowe'en falls precisely on October 31st, but formerly it would have been celebrated at the full moon closest to this time since outdoor meetings were made easier by the light of a full moon.

What do all the trappings of Hallowe'en mean? The turnips and pumpkins, the masks and disguises?

It's harvest-time for turnips and swedes. If you found a really big turnip you could hollow it out and make a mask with big jagged teeth to frighten evil spirits away. These spirits might have been seen as rats that threatened the grain store, or as diseases that might attack the cattle which were being brought in for winter. Pumpkins were a more recent introduction from America, but they're used in the same way.

Masks and disguises would have been worn to confuse any of the harmful spirits that might have been around, like the ones that whistle through the trees in autumn, tearing all the leaves off and bringing bad luck. If you looked like someone else they wouldn't recognise you. We see this at Lewes on Bonfire Night, with its legions of Vikings, cavaliers, Zulus, pirates and so on. But, of course, these days it's mostly just for the fun of it.

Why are witches associated with Hallowe'en?

It's one of our main festivals. It's as if the different levels of existence are much closer to one another at this time. The veil between the world of ghosts and spirits and the world of people is much thinner, so the wisdom of these spirits is available

Do you think that the commercialisation of Hallowe'en, with its model witches, trick webs and rubber bats and toads, dilutes the potency of the event?

I think this is rather a good thing. It's better to laugh at something that you're not sure about, than to be afraid of it. Witches got burnt in the past because people were scared of their power. In becoming familiar with witchcraft, even through the jokeshop, people might be encouraged to ask deeper questions.

How do you respond to people who dismiss ghosts and spirits as so much claptrap?

Witches don't preach, on the whole. If someone doesn't accept something, I think they're entitled to their view. But I have experienced these things. I have felt the presence of spirits of departed people.

What's the difference between Hallowe'en and Guy Fawkes Night? They seem to come very close together.

If we leave aside the more recent addition of Guy Fawkes and the Gunpowder Plot, they're both really part of the same festival. We have bonfires at Hallowe'en as well as on Bonfire Night. The underlying theme that unites them is the attempt to stop the sun from going away.

These days, with central heating and city life, the departing sun doesn't mean so much to us. But in the past, the cold, dark winters weren't much to look forward to.

Ancient people didn't need modern science to tell them that the days were getting shorter, or that the sun was a great big bonfire burning up in the sky. So they used what skill they had to light bonfires and send up firecrackers, to preserve the warmth of the sun by imitating it.

Why are effigies burnt at this time?

Traditionally the King of Winter, who fought and beat the King of Summer at harvest-time, was burnt in effigy on the bonfire at Hallowe'en so that his spirit would be released up into the sky for the sun. It was a sacrifice to encourage the sun's return.

There is a dark, cruel, wild side to human nature. This is what the old pagan religions sought to bring out and make part of their festivals. If you wanted to hate, burn and destroy, there was a time and a place for this. Bonfire Night at Lewes, with all its complexity, is a survival of this. At Hallowe'en, traditionally, you brought symbols of your failings and mistakes and threw them on to the fire. You destroyed the things that you hated about yourself.

The desire to destroy things is part of the dark side of human nature, just as the singing and dancing at Padstow in May is part of the light side. In this sense, Padstow and Lewes are almost opposite in their meaning. Padstow celebrates life overcoming death in the daylight, whereas in Lewes effigies are destroyed at night amidst the symbolism of death and destruction.

Remembrance Day is about death. Is there any significance to its falling at this time of year?

Well, the First World War ended at this time of year – the armistice was signed on November 11th – so there's no immediately obvious significance. It seems more of a coincidence. But the poppies are interesting. They grew in the fields of Flanders and we wear them as a reminder of the hundreds of thousands who died in battle.

In pagan celebrations, garlands of poppies were picked and used with blue cornflowers to deck the horses and carts of newly gathered corn. Around this 'bier', carrying the spirit of the King of the Corn, the people would have walked wailing, or in silence, mourning the spirit of the King of the Corn cut down in his prime to provide nourishment. Similarly, we mourn the young men sacrificed in war, so that others may reap a new harvest.

Perhaps lingering memory drew together the ending of the farming or Celtic year, with all its death symbolism, and the date of Armistice or Remembrance Day. By this time the poppies are long dead in the fields, yet their blood-red petals at the foot of memorial crosses and in our lapels recall the sacrifice.

We can see the lines 'At the going down of the sun and in the morning, we will remember them' (Laurence Binyon) as referring to both the slain youth and the reaped ears of corn which by now are winnowed and safely stored for winter.

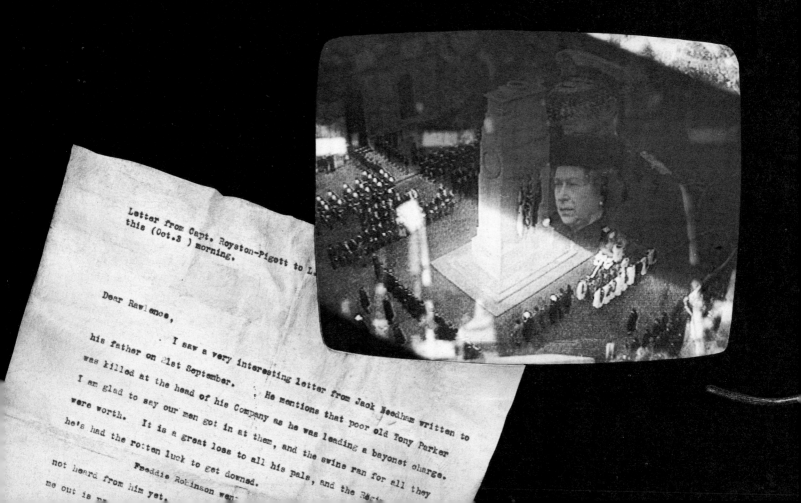

October 28th, 1984. This morning the radio announced that British Summer Time had ended at 2.00 a.m., which means it will be getting dark by late afternoon. Then, on the news, came another confident pronouncement by the government that coal stocks would last through the winter and the miners' strike would be defeated.

In spite of the government's hopes, we're already stocking up with candles and paraffin in anticipation of power cuts. The realities of a winter without enough electricity are looming. Secretly, I'm quite looking forward to the cosiness of living by candlelight.

A change will come over the local newsagent's shop this week as well. The glass cabinet that usually houses birthday cake candles and pencil sharpeners will be stocked with brightly-coloured boxes of Vesuviuses, Catherine wheels and Roman candles. On the streets outside, striking miners collecting for their winter hardship fund will be joined by hordes of children, wheeling stuffed jumpers in rickety baby buggies. As Bonfire Night approaches I'll be repeatedly asked for 'a penny for the guy' in one of the few remaining pockets of legitimate begging.

BOB BUSHAWAY – BONFIRE RITES

This moment of the year marks the transition from summer and early autumn, where harvest was the significant event, into the depths of midwinter. A century ago we'd have been looking back in a very real sense to the economic plenty of harvest, where chances of full employment were more certain and wages were better. And we'd be entering a much more uncertain period where employment was less dependable.

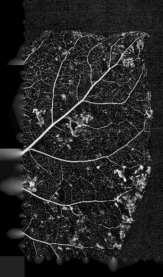

Why is fire such a central part of festivals at this time of year? Particularly on Guy Fawkes Night.

There are two ways of looking at these fires. One is from the anthropological point of view, where they are seen as a kind of sympathetic magic having pagan origins, the purpose of which was to invoke the sun's dying heat for winter. The other view, which I'm more concerned with, is that in former years wood was an essential fuel. Unless you lived in those areas where there were natural outcrops of coal, wood was needed for cooking and heating. People used to take dead wood from the hedges and woods of the countryside around them. They saw it as a natural flotsam that was free rather than having any right of property attached to it. But this custom was often hotly contested by landowners and local lords of the manor who saw the wood as their own. So at this time of year, when the dead wood is revealed and winter is approaching, people gathered dead wood and made bonfires as an affirmation of their right to collect and burn that dead wood freely at all other times of the year. It was a way of reinforcing a genuine economic custom.

Why do they burn the Pope instead of Guy Fawkes on Bonfire Night at Lewes? Who are the Lewes martyrs?

During the reign of Queen Mary I Protestants were persecuted as an attempt was made to restore the Roman Catholic church as the state religion. At Lewes in 1557 seventeen Protestants were executed by burning.

Why burn effigies?

Bonfire Night has often been used as an occasion for social criticism which might not have had an outlet at other times of year. If you burn someone in effigy you stigmatise them. This often happened to local people who'd offended popular morality; for example, the local squire, for taking a stand against the church choir.

But at Lewes we saw Neil Kinnock exploded and Michael Foot being blown up on a cruise missile. And you got a banger in your back. Don't you feel politically ambivalent about all that?

Yes. It was anti-C.N.D., which perhaps reflects some local feeling this year. This is depressing, but they also played 'The Red Flag' along with 'Rule, Britannia!'. And in the past it's been the turn of others: Margaret Thatcher last year, Idi Amin and the Ayatollah earlier. Of course it is important to remember that in burning an effigy, the crowd are not actually burning real people. The Pope is burnt in effigy at Lewes rather than any of his adherents in reality. Effigy burning can both dissipate actual hatred, or contrastingly, focus aggression and increase prejudice. It is a powerful form of ritual not to be treated lightly.

This whole ambivalence was summed up for me at Lewes when I met someone taking part in the Cliffe bonfire dressed as an S.S. storm-trooper. He told me that he'd intended coming as Lenin but couldn't get the costume together in time. When I said, 'I see you've settled for the totalitarianism of the right rather than that of the left,' he said, 'Actually, I'm a Liberal and I've just come from a Liberal party selection meeting.'

As much as anything else, people were having a boisterous and rowdy time. They were enjoying themselves, and affirming the identity of their local community.

In the past there's been a lot of violence here, between the town authorities and the local people. The authorities felt that the annual procession with its flaming tar barrels, bonfires and burnings of the Pope were a very real fire hazard to the town which in those days was built mostly of wood. From the eighteenth century onwards several attempts were made to suppress the custom but each attempt ended in riots; violence was the result of people trying to defend their custom.

The climax came in 1847 when the Metropolitan Police were called down from London to suppress the celebrations once and for all. Without success: the result was an even bigger riot. Shortly after that a compromise was reached: the event would be allowed to go

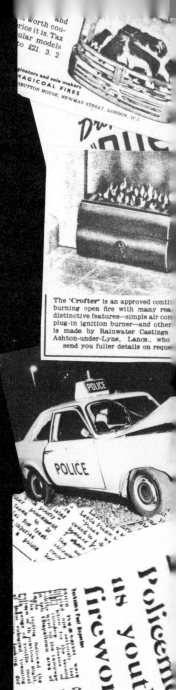

unmolested if the people controlled it better. That's how the Bonfire Societies came about. they organise the occasion.

Is there a pattern to the way the Victorians suppressed popular customs?

Broadly speaking, yes. They suppressed them, remodelled them and handed them back transformed. It's as if they were saying: 'Here you are. Now you get on with this. It's much better than the sort of rough and tumble you used to go in for. We won't get the same number of injuries and pregnancies in the village as before.'

Whereas in actuality they were acting against anything that showed working people in a role of independence, taking initiatives beyond the control of the local authority – the church or landed gentry. And they were challenging anything that overturned property rights, such as the use of the streets in Lewes. Lewes and Padstow are among the few customary events that survived this kind of suppression in the nineteenth century in one piece.

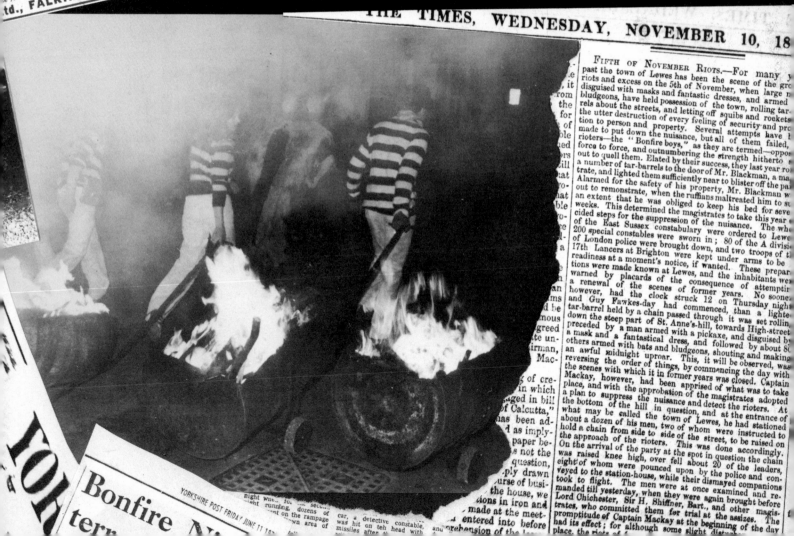

THE TIMES, WEDNESDAY, NOVEMBER 10, 18

FIFTH OF NOVEMBER RIOTS.—For many y
past the town of Lewes has been the scene of the gr
riots and excess on the 5th of November, when large n
disguised with masks and fantastic dresses, and armed
bludgeons, have held possession of the town, rolling tar-
rels about the streets, and letting off squibs and rockets
the utter destruction of every feeling of security and pro
tion to person and property. Several attempts have b
made to put down the nuisance, but all of them failed,
rioters—the "Bonfire boys," as they are termed—oppo
force to force, and outnumbering the strength hitherto
out to quell them. Elated by their success, they last year ro
a number of tar-barrels to the door of Mr. Blackman, a ma
trate, and lighted them sufficiently near to blister off the pai
Alarmed for the safety of his property, Mr. Blackman w
out to remonstrate, when the ruffians maltreated him to su
an extent that he was obliged to keep his bed for seve
weeks. This determined the magistrates to take this year
cided steps for the suppression of the nuisance. The wh
of the East Sussex constabulary were ordered to Lewe
200 special constables were sworn in; 80 of the A divisi
of London police were brought down, and two troops of t
17th Lancers at Brighton were kept under arms to be
readiness at a moment's notice, if wanted. These prepar
tions were made known at Lewes, and the inhabitants wer
warned by placards of the consequence of attemptir
a renewal of the scenes of former years. No soone.
however, had the clock struck 12 on Thursday night
and Guy Fawkes-day had commenced, than a lighte
tar-barrel held by a chain passed through it was set rollin.
down the steep part of St. Anne's-hill, towards High-street
preceded by a man armed with a pickaxe, and disguised b
a mask and a fantastical dress, and followed by about 80
others armed with bats and bludgeons, shouting and making
an awful midnight uproar. This, it will be observed, was
reversing the order of things, by commencing the day with
the scenes with which it in former years was closed. Captain
Mackay, however, had been apprised of what was to take
place, and with the approbation of the magistrates adopted
a plan to suppress the nuisance and detect the rioters. At
what may be called the town of Lewes, he had stationed
about a dozen of his men, two of whom were instructed to
hold a chain from side to side of the street, to be raised on
the approach of the rioters. This was done accordingly.
On the arrival of the party at the spot in question the chain
was raised knee high, over fell about 20 of the leaders,
eight of whom were pounced upon by the police and con-
veyed to the station-house, while their dismayed companions
took to flight. The men were at once examined and re-
manded till yesterday, when they were again brought before
Lord Chichester, Sir H. Shiffner, Bart., and other magis
trates, who committed them for trial at the assizes. The
promptitude of Captain Mackay at the beginning of the day
had its effect; for although some slight distur

Bonfire N

YORKSHIRE POST FRIDAY JUNE 11

The smell of the morning after is the strongest memory. It would hang heavy in wet mist – the remnants of the bonfire blended with the soot of spent rockets and the scent of liquid leaves that we'd trampled underfoot as we roasted round the flames. It was a sad smell too, because the next day was so anticlimactic. For weeks there'd been the excitement of choosing fireworks, making the guy and building the bonfire, leading up to the darkness, the baked potatoes, the whooshes, bangs, flashes and the lightings with gloved hands. Then nothing but the smell and a long grey stretch until Christmas.

The Lewes Festival in November is near the start of the old Celtic year, which begins in darkness. It's close to Hallowe'en and All Souls' Eve, and it seems like a procession of the ghosts of the past. It's as though the dead are coming back as we enter the night of winter, because everyone is dressed up as figures of the past, including ghosts.

But the bonfires show that even though the sun is going, we have the ability to make fire and keep ourselves warm.

It seemed as if the dressing-up was giving us successive layers of British history, which were falling all over the place in a very incoherent, dreamlike association, like a kind of collective unconscious of Britishness.

Yes. And you go down through similar strata in dreaming, from the personal to the collective, but Lewes is still very confused, still wondering about itself, still evolving. Unlike Padstow which, as we've said, was a very deep, still dream, Lewes seemed like a kind of half-way house where there is both time and non-time.

What did you think about the underlying violence at Lewes? The blowing up and burning of effigies?

Perhaps it's better to burn effigies than actual people. It's as though the festival gives people a chance to purify themselves by getting rid of their anger.

Don't you feel there's a difference between burning effigies of living politicians, like Michael Foot and Neil Kinnock, and burning effigies of dead people, whether it's Guy Fawkes or the Pope?

I didn't much care for the blowing up of Foot and Kinnock. This seems to be doing nothing but raising anger, raising up spirits against the living.

But the burning of the sixteenth-century Pope seemed more powerful. It was more like conversation. If you burn effigies of dead people, you're bringing them back for what they can give you, then you send them away again. That is, in psychological terms, you are conversing with a part of yourself that is not conscious in order to relate to it.

EVENT 2958. 1279.

Everybody likes to believe that they are pure and shining white and they don't want to be nasty to anyone. This attitude of course casts a very dark shadow in front of you. If you pretend that the shadow doesn't exist and that you're incapable of horrible acts, these capacities are still in you but they're going to work unconsciously. But if you relate to your bestiality, if you converse with your shadow and know the horror within you, then it is less likely to go out of control. If you know your dark side you have a fair chance of being human.

Burning the Pope at Lewes perhaps is a way of conversing with our shadow. The basis of the Christian religion is similar. We bring back our God every year to be killed again. God is crucified and this is the worst thing that can possibly be done. Yet he resurrects and we are forgiven. Likewise, the Oss is brought back year after year to dance, die and resurrect.

Now, if the hatred is personal, like the burning of Foot and Kinnock, there may be no 'resurrection'. But if the images are ritualised, as they are with Jesus, Guy Fawkes and the sixteenth-century Pope, the whole thing is seen as trans-personal, and the energies that are released in us are healthy.

Is there some resistance to outsiders coming to festivals?

I'm sure there is at Padstow, and should be. These things can be so easily de-natured, spoilt and lost.

And if we're pointing a camera at them ...

... I can't speak for them, but they might feel that you were trying to steal the soul of the Oss.

Do you think we would be?

You might be doing yourselves a disfavour by putting the experience just into a camera. You might be too concerned with your filming to participate.

As the active participation in customary events has progressively decreased, an anxious voyeurism has taken its place. These days most of us just look on as tourists, and try to ignore the nagging restlessness of not belonging. Perhaps, deep down, we miss the affirmation of being a part of a community that only the local participants in these events can experience. Perhaps, deep down, as outsiders to these rites of passage, we experience a gnawing sense of loss.

Whatever the diagnosis, certainly the manic click, click, click of camera shutters indicates that something is wrong. It's as if the ceaseless taking of photographs anaesthetises anxiety – by reducing the sacred time of the ritual to a one-eyed observation through a view-finder. In this process, the sense of collective identity that events like Padstow generate becomes little more than an agitated search for the best viewpoint, a quest in which the observer is all too often insensitive to how much the camera disturbs the observed.

BOB BUSHAWAY – CAMERAS AND FESTIVALS

Do you think the presence of our camera has any effect at these festivals?

The great late-nineteenth- and early-twentieth-century folklorists thought that if they didn't note down complicated festivals in meticulous detail, the traditions wouldn't survive, the customs would be lost for ever. The contradiction was that although they thought they were saving them they were in fact partially destroying them by fixing them in time, like the proverbial fly in amber. By recording them they undermined the oral continuity of the customs.

Obviously the act of memory – which verse comes next, which way do we turn? – is very important to this continuity. But lapses of memory are also very much part of the developing process of a custom. If you can't remember whether you turned right or left, the fact that you might plough straight on through someone's house is a change resulting from a lapse of memory!

It's a bit like those strange verses in the Padstow song which seem to make no sense.

Yes. It's the agglomerations of mistakes and additions through lapses in memory which give life and vitality to the event for the local community. It happens in songs when people can't quite remember the words and improvise spontaneously. There's also a lovely example of it in the mummers' plays. One of the crucial characters is known as the 'Turkish Knight' in many versions, but in one surviving Hampshire version 'Turkish Knight' clearly made no sense because it has become 'Turkey Snipe', which perhaps means much more to a Hampshireman.

Now, the act of recording or filming might prevent these changes happening by fixing the event. So the custom might become stagnant.

Do you think this explains the hostility to film crews at the Padstow Festival? At one point we were told to pack up our equipment in no uncertain terms.

I think you've got to be very careful how you try to record an event. A camera crew can easily affect a festival, either through drawing the hostility of the crowd, or by seeking to alter the event in some way. Occasionally I've witnessed the horror of the 'retake' at festivals. You know: 'could you possibly walk down the street again?' This kind of intrusion into the event is almost criminal. But if you record very carefully, using the 'fly on the wall' technique, you might 'capture' the event without disturbing its significance.

A much larger intrusion of the media into customary events arises through the media's development of national stereotypes. Christmas is a good example of this. In the eighteenth century it would have been very much a local affair, with visits of wassailers and mummers, who would have been known to the local people. And often they would follow the same routes taken at Maytime or Harvest Home. But with the rise of the popular press a stereotype was fixed. With Christmas you can almost point to the historical moment. There was a famous cover of the *Illustrated London News* which depicted Victoria and Albert celebrating Christmas at Windsor, surrounded by children, presents, tree and turkey.

And this domestic image, of Christmas in the bosom of the family, was selected in preference to the local experience. Wassailers and visitants became an intrusion. In this way a stereotype developed which fixed the morality of the custom, making it more bland, and destroying its local basis.

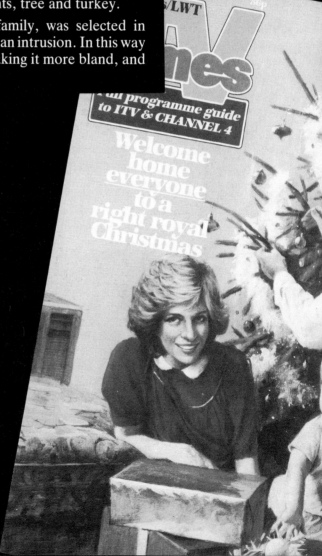

MARIAN GREEN – CHRISTMAS

What are the origins of Christmas?

Christmas was a festival which the Church imposed on the ancient pagan midwinter festival of Yuletide.

This used to be a time of fear: the sun had retreated and the days had got so short that it almost seemed that they were disappearing.

Now, those who observed the exact movement of the sun could see that it reached its lowest midday point on about December 21st. This is the winter solstice, when the sun appears to stand still. But a few days later, the daylight would have lengthened just enough for people to realise that the sun was returning. This would have been about December 25th and would have been a time of great joy.

So while we think that we're celebrating Jesus's birthday on December 25th, this child in the manger has come down to us from the ancient festival where he was the Starchild. The star is a symbol of hope, brightness and promise that goes back to pagan times. The Starchild represents the power of the great sky god, the sun, returning among the people.

Jesus was born in a stable but sometimes this is shown as a cave. The Roman god, Mithras, had his birthday on December 25th and was also born in a cave. He was worshipped all over the Roman Empire just before the advent of Christianity. These caves are the womb of the earth goddess, from which life springs at the dead of year.

Why do we hang up decorations at Christmas-time?

The good old pagan belief in life overcoming death underlies much of the concept of Christmas, and you'll find it if you dig down through the plastic and glitter of today's mass-produced artificial decorations.

When people hang up decorations today they're basically trying to escape winter and pretend that it's summer.

Prince Albert is credited with having introduced the Christmas tree to England in the last century, but in Scandinavia they've been decorating trees to symbolise the sun-god for thousands of years.

Holly and mistletoe are pre-Christian symbols of life at a time of death. When all other trees have shed their leaves and appear to be dead, the holly and mistletoe trees are not only green but are bearing fruit.

The hollybush, with his red berries the colour of life and his prickly leaves, is king of the forest. The ivy, who entwines, is a symbol of the goddess: she who clings to you will keep you upright.

Mistletoe was sacred to the Druids. Its white berries, ripening at the dead of year, were thought to be the seed of the sun-god. So when men and women kissed under the mistletoe, the man would eat one berry for each kiss in order to receive the sun-god's virility.

The Christian church didn't like mistletoe because of these pagan and sexual associations. 95

Good King Wenceslas looked out
On the feast of Stephen
When the snow lay round about
Deep and crisp and even
Brightly shone the moon that night
Though the frost was cruel
When the poor man came in sight
Gathering winter fuel.

THE SUNDAY TIMES

2 JANUARY 1983

No 8269 Price 40p

New Year's deadly night of

Two die
MPs

Striker killed digging for coal

By Malcolm Pithers

A 37-year-old striking mineworker died under tons of coal and slurry yesterday when he was digging collapsed on top of him.

Mr Knapper, of Weelan near Wake-

stances since the miners' dispute began. A 14-year-old boy was killed a few weeks ago near Upton colliery, Pontefract.

Mr Knapper had been on strike throughout the dispute and was collecting coal to burn at home. The area where the accident happened is known as Pineapple Hill, Normanton, not far from Knapper's home. Local miners

Extremely heav Yorkshire at the loosened much of

The vicar of Reverand Eric the entire vill shattered by ne just digging as fires at home to sermonise Mr sion but it of him not small miners for their something and if I had in re position.

MOONSHINE

You could wade knee-deep out to the fort along the sand-bar at the low low tides when the moon had done its work. Some afternoons we'd shrimp the whole way there, stay too long, and hurry back waist-deep, in rising anxiety.

Once on dry land, we'd run off towards Seaview with our nets and little bags of pink wet shrimps. Johnny always got back first, because when we reached the pebbles I'd slow to a barefoot hobble, while he skipped over the hard round stones as if they were still sand.

One day, when we were still some way from the town, he stopped at the Ladies. It was a storm-worn, red-brick amenity standing all on its own, and there was Johnny, daring us both to go in. And did I have a shilling? Because he knew we could get towels inside.

There was no one around as we crept on to the sandy cement floor into that place, with its cream gloss walls.

How dry and clean it is. But we shouldn't be here, nearly naked, whispering in case anyone's behind those closed doors. I want to get out as I pass him the shilling from the neat pocket of my trunks. I don't want to be here as he turns with a smirk to the sweet machine on the wall, stretching on tiptoe to drop the coin in the slot. But he pulls open the wooden drawer with the ease of the one who dared, and I know I must stay because there's no going back.

Who'd have thought they sold towels in a box, like Payne's Poppets. And so small. How can you dry your-self on that? And why pins? Won't they prick? I know it's no ordinary towel. Johnny knows what it's for. His tone says it's for something down there that's not us. He knows and he's not telling me. I think it's to do with his big sister Sarah, who has a room to herself and gets dressed on her own this holiday.

Well he doesn't quite know, but it's rude. Johnny grins as he lets me in on the secret he hasn't yet grasped. It's for what happens when girls stop being girls. Grinning back boy to boy, I quiver with sexual complicity, on the threshold of knowing everything without being one bit wiser.

It was cold in there. I didn't want to be caught. I wanted to throw it away, hide it, run out into the sunshine.

This morning there's a gibbous moon waning up in the blue sky over the city. The first jets are just grumbling in from their long hauls through the night. One of the earliest facts I learnt about the moon was that it reflected the light of the sun and that's how we saw it. But I've always been baffled by its changing shape. For example, when it's new, is the crisp inner curve of its crescent formed by the earth's shadow? No, I'm wrong – that's an eclipse. Besides, the blurred terminator of a three-quarter moon couldn't be the earth's shadow. It curves the wrong way.

Today, however, I can see how the shape of the moon which is in front of me, is defined by the sunlight coming from behind me. All is suddenly clear: when both sun and moon are in the sky together, the shadow makes sense.

'I've accumulated an inpenetrable welter of facts about the moon over the years. For example, it has a magnitude of −12.7 at maximum illumination, when it has an observed angular diameter of 31.09′; it has an orbital eccentricity of 0.043–0.0668 and an inclination of 5.1°; it has a radius of 1.738×10^8 cm and a mass of 7.35×10^{22} kg; $m/(4/3 \, \pi r^3)$ produces a mean density of 3.34 g/cm^3. From the surface of the earth, 500,000 craters can be seen . . .

Gradually this kind of factual wadding has filled the hollow left inside me in the Ladies that afternoon by Johnny's mysteriously incomplete answer.

99

Why has the moon been associated with women in so many cultures?

The moon measures out in the heavens four phases of seven days each which make a lunar month. This regularity can be observed by anyone who looks up at the sky and it must have been one of mankind's first time-keepers. What must also have been observed from earliest times is the connection of this measure with human fertility: the lunar cycle and a woman's average menstrual cycle are both twenty-eight days long. In fact early moon calendars have been found with intervals marked with menstrual blood, which was the sign of fertility.

The moon isn't the only thing in nature that we respond to. We respond to the sun and the seasons, day and night. We have many inner clocks, but one of these in women does seem to be a moon clock. This connection has been much disputed, but it seems likely that the womb responds to the moon as it changes the earth's electro-magnetic field in its monthly orbit of approach and retreat. This is the weather of the universe.

In ancient times this would have seemed magical: not only did women give birth to children, but they had this connection with the moon in the skies. They became like a walking moon on earth.

Do you see this as magical?

Yes. A woman carries round in herself this extraordinary power that men don't have: the capacity to have children. This is the first and original magic: without it there would be no people, and, for us, no world.

This power of childbearing is signalled by menstruation. The relationship of menstruation to conception and childbearing is traditionally this: a woman produces an egg; if it's fertilised it can grow into a complete human being; if it is not, it can release the energy

that would have made the human being into the woman's body for her own use.

This is the traditional view and it seems to be close to the truth.

What has happened in our culture is that the exquisite sensitivity and charging of the body that occurs at the menstrual time has been shut away from the woman herself.

And men have done the shutting away?

Yes. By ignoring it, treating it as an illness, as madness, or as the 'curse'.

Why?

Throughout recorded history men have been jealous of any power, and women are born with this power. Anything full of power in this way is dangerous, which is why men have made it taboo.

We find the menstrual taboo buried in fairy tales, and in some it gets very beautiful and direct treatment. One of the clearest is 'The Sleeping Beauty'.

You will remember that twelve fairies were invited by the king to the christening of his daughter, the princess. But there was also a thirteenth, the wicked fairy, who was not invited. So she put a 'curse' on the child: that on her tenth birthday the princess would prick her finger on a distaff and bleed. And everyone in the castle would fall asleep for a hundred years.

So to avoid the curse the king ordered all the spinning wheels in the land to be destroyed.

But on her tenth birthday the princess entered a secret chamber in the castle where she discovered an old woman spinning. And of course she pricked her finger, which bled. Everyone fell asleep for a hundred years while a high wall of thorns grew up around the castle. Then a magical prince cut his way through the thorns and woke the princess with a kiss.

Bruno Bettleheim, in his *Uses of Enchantment*, shows how menstruation is the female, or fairy's 'curse'; and it is plain that had the king not tried to keep the princess from puberty, the bleeding would have been no curse at all. He was denying her her rite of passage. Why thirteen fairies? Because the solar, father's year, is divided into twelve months, while the woman's year of experience is divided into thirteen periods, since twenty-eight days is the traditional length of the menstrual cycle. To forget this, to forget the thirteenth fairy, is to bring on a curse. Bettleheim believes that:

'The story of Sleeping Beauty impresses on every child that a traumatic event, such as a girl's bleeding at the beginning of puberty, and later, in first intercourse, does have the happiest consequences. The story implants the idea that such events must be taken very seriously, but that one need not be afraid of them. The 'curse' is a blessing in disguise.'

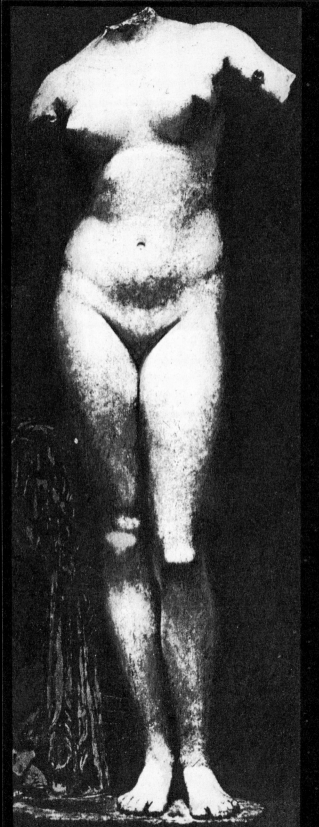

What are rites of passage?

A rite of passage is a way of travelling from one condition of life to another without losing consciousness.

Festivals are rites of passage that initiate us through thresholds of the year: from spring to summer at Padstow, or summer to winter at Lewes. But there are also rites of passage for the initiation of a person into a new stage of life. For example, in some cultures, the young woman who starts to menstruate is initiated into her new capacity to become a mother, if she is to bear children, or a shaman or magician if she is not. Unfortunately our society has no such initiations, or rites of passage, for either men or women, so we enter new stages of life with no image of our capacities, save those provided for us by the mass media.

Similarly, there is no sense of threshold, no understanding of what a person may experience symbolically in a ritual for them to enter the menopause.

What are the consequences for us of having no rites of passage?

There will always be an interior initiation. A person approaching a crisis or a threshold in their life will have dreams which symbolically reveal possibilities and ideas for conduct. If there is no ritual there will still be these experiences spontaneously in dreams. I believe it's necessary to encourage and amplify these dreams, to converse with them. If we study our dreams we provide these rites of passage for ourselves.

What about women who don't experience menstruation positively?

A lot of women don't, and perhaps this is the result of the shutting away, that it becomes painful instead. We don't yet know. It's only just now, when women are beginning to share their experience in this neglected quarter, that the real answers are emerging. One of these is that menstruation is a time of positive resource rather than an illness.

What do you think men can learn from menstruation?

The menstrual cycle is a 'feeling' thing, but men don't see this because we live in a masculine non-feeling culture. Feelings are left to the women to exercise.

Yet the most important thing for anybody to understand is that what happens inside another person is a reality, however different it may be from your own reality.

The events of the menstrual cycle are among the foremost things that happen to people. In a household where a woman has a powerful menstrual cycle, there's a bodily, sexual and mental rhythm set up between the two poles of ovulation and menstruation.

I believe that men can gain great enhancement in their relationships with themselves and with women by acknowledging and attending to this rhythm instead of ignoring it. By seeing what actually happens.

What is your own experience here?

Penelope, whom I live with, no longer suffers from menstrual distress. In the premenstrual time she now becomes full of an extraordinary imaginative and sexual energy. She experiences upsurges of energy which do not distress her any more. And she uses these creatively.

But how does this affect you?

I have these cycles of dreams which come to a creative resolution actually at the period. During the premenstrual week I have very strong inchoate creative impulses which are rehearsed in my dreams. I can start a piece of creative work, say a play or some psychological work, at the beginning of the month, and the solution will come with Penelope's period. Our daughter Zoe also has these dreams which I can describe as a lability of feeling in the premenstrual week.

Don't you think that many women would object to being defined simply by their childbearing capacities?

Yes. And it's something that the world's great religions have done. But I am not saying this. All I am saying is that the womb is an extremely powerful and magical organ and it's foolish to ignore these capacities. A woman's womb gives her two sides: ovulation, when she has the possibility of being the person who conceives a child, and menstruation, when she can be her own self, using her energy for imaginative and creative tasks, or withdrawing to the quietness of contemplation (a special 'sacred' time which in our culture there is no provision for).

This dialectic runs through interior life. It is the pattern for our humanity, yet it's something which the male culture has lost.

What about men who don't have intimate relationships with women? Do gay men somehow bypass the menstrual taboo and all its problems?

Every man has a relationship with at least one woman and that's his mother. A man can find perhaps that something cyclical in his nature has been established in him in childhood. Something which comes and goes in his energy which resembles the menstrual rhythm, and is perhaps picked up from his mother's attitude to her menstruation. Now if his mother, as is all too frequent, hated herself because she had periods which she'd been taught were dirty, nasty and taboo, then this hatred and suspicion would be passed on to the boy and the man.

There is something which psychology has hardly touched yet which we might call the 'menstrual trauma' in men. When his mother has her periods she may be full of emotions that are strongly loving and energetic but which have no form of expression because of the taboo of menstruation not being allowed for in the family. For the boy, who is representative of masculine culture, this is a great puzzle: he is both feared and rejected as representative of the culture that taboos menstruation, but he is loved and desired as the child in the flux of emotions at this powerful 'feeling' time. This contradiction is the menstrual trauma which boys pass through but don't understand. It becomes something about one's childhood history which is concealed.

My mother suffered most terribly from her periods. I remember how the whole house would be charged with an awful anger and I didn't know why this was. Why was my father alienated? Why did my mother knock things over and swear, saying, 'Hell's Bells and Buckets of Blood!' She had these energies in her anger but she treated her period as if it were a mad, awful thing not to be spoken about. It was something about her that she didn't like but which was important to her. And she didn't like it because she'd been brought up not to like it. I would ask why she felt like that and she'd say, 'It's a woman's time of the month.' But what was the meaning of this special vehemence she was filled with?

And then, when she calmed from her rages, she would tell the most marvellous stories, which I now think came from her dreams. And this has been found by women who have made studies of menstruation: that it's a special time for the imagination, for discovering images which are important and life-giving.

Do you think you've missed something in not being a woman?

I think so, yes. The direct experience of growing a person inside yourself must be extraordinary, and a man can only have a shadow of this. This is moon envy! Yes, I think that if I could have a child I would.

You seem to be suggesting that men can only get access to certain of life's fundamental energies and resources in a secondary fashion, as it were, through the mediation and example of women. Isn't there a danger of confining women to the traditionally feminine through a kind of idealisation? While at the same time men become some kind of inferior beings?

Yes and no. Men and women are all faced with the problem of

gender roles. When we're children it's as if we are both sexes, but as we move towards puberty, we divide, becoming a man or a woman.

But that doesn't mean to say the rest is lost. It goes underground and becomes an unconscious counter-sexual figure. For the woman there is a male dream person in her, which Jungians call her animus. And for the man, a female dream person, his anima. So the same person is both masculine and feminine, and needs to find the balance.

Many men have a strong feminine moon side, and they can discover this when they see that their dreaming patterns are influenced by the moon. But many men do not, and to get this magical capacity they have to learn from women.

The experience of women is that they return again and again to this life necessity which is required of them by their wombs. And menstruation is such a strong thing that it's too much for many women to deal with, especially when they are given no image in our culture to deal with it. Whereas women seem to be more able to transmit life's energies, men are more like a reflecting mirror to existence.

Do you think that men experience time in a more linear way through not experiencing the menstrual cycle?

Yes. Men tend to think as the sun thinks, in solar time. The sun doesn't change or alter its shape. It simply rises, goes across the heavens, and sets. The intellectual masculine culture is a solar culture in that it thinks only in the light. But the moon ebbs and flows. It is both light and darkness. In letting the dark in it's an alternative to the culture we have.

Men's time seems to be in straight lines, directed towards an object or a project. The great example is the Christian example, which I think is the man's view: the whole world started with Genesis; we had original sin in the Garden of Eden, from which point we go straight through to the Last Judgment where everything is settled.

Whereas for women, time is more a spiral, a re-circling. I was very impressed by the image of the spider's web that the women put on the wire at Greenham Common. The spider is something which people are afraid of, an image of catching and devouring, but the Greenham women have made this image of a web which is not just straight lines, but the covering of a view.

Thought moves in time, and the way we think embodies how we think about time. The Greenham women say we will go from this point to that point, seeing what is important, going here, going there, and then we will knit it together in a spiral. It is an interaction, a hologram.

There was also a slogan on the wire at Greenham which read: 'Warfare is disguised Menstruation', and it seems to be the case that the cultures that wage war most strongly are those which have the strongest taboos against menstruation. Our culture, which is the most bellicose in history, has either a very violent language of slang words against menstruation, or complete silence. On the other hand, there are cultures which honour women more, and which don't have such strong sexual and menstrual taboos. And these tend to be more peaceful and less likely to wage war.

Why do you think that the women of Greenham Common dressed as witches at Hallowe'en?

MARIAN GREEN: They were probably trying to symbolise the power of the female. For so long now the innate psychic power of women has been frowned upon by men. Their understanding of things which logical men couldn't see has always been feared and often persecuted. The women at Greenham are putting forward a very important case – the desire for peace, fertility and continuity.

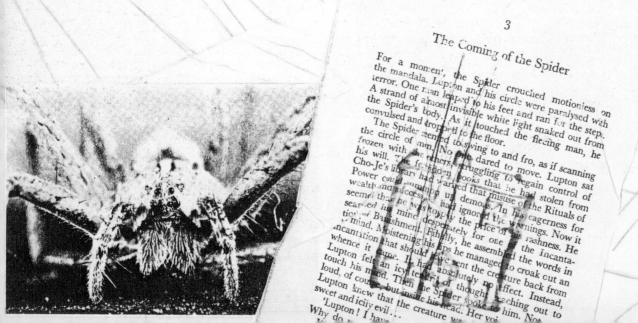

3

The Coming of the Spider

For a moment, the Spider crouched motionless on the mandala. Lupton and his circle were paralysed with terror. One man leaped to his feet and ran for the step. A strand of almost invisible white light snaked out from the Spider's body. As it touched the fleeing man, he convulsed and dropped to the floor.

The Spider seemed to swing to and fro, as if scanning the circle of men. No one dared to move. Lupton sat frozen with the others, struggling to regain control of his will. The forbidden books that he had stolen from Cho-Je's library had warned that misuse of the Rituals of Power could summon up demons. In his eagerness for wealth and success, he had ignored the warnings. Now it seemed that he was to pay the price of his rashness. Finally, he managed to croak out an incantation that should send the creature back from whence it came. He had absolutely no effect. Instead, Lupton felt an icy tendril of thought reaching out to touch his mind. Then the Spider spoke inside his head. Her voice was loud, of course, but inside his head. Her voice was sweet and icily evil . . .

'Lupton! I have . . .'

JUDITH HIGGINBOTTOM: The spider's web became a very important symbol of what was going on at Greenham Common. When the spider makes its web it weaves around in a spiral rather than cutting through in a straight line. This seems to be very much the way women organise and do things. Arachne, the spider, is to do with intuition, which this society sees as a particular quality of women, perhaps because men don't use their intuition very much any more.

107

The search for proof that astrology does or doesn't work is a contemporary obsession. Equally tempting is the desire to prove the physical effect of the moon on menstruation. Yet attempts to authenticate an essentially shared experience with the stamp of experimental evidence somehow devalues the subjective: apparently objective science is called on to arbitrate over the immeasurable intimacy of human inner life and emotion.

In her project on menstruation, 'Water Into Wine', Judith Higginbottom was looking more for patterns than proof, and affirmation of what had been for her an isolated and difficult experience.

JUDITH HIGGINBOTTOM is an artist and writer. She lectures at Falmouth School of Art, and was European Fellow in Fine Art at Exeter College of Art and Design and and the Studel Academy, Frankfurt. She has held several video and film screenings of her own work throughout the world, exhibiting in New York, Tel Aviv, London and Liverpool.

JUDITH HIGGINBOTTOM – TIME, MENSTRUAL DREAMS AND THE IMAGINATION

What led you to start exploring the menstrual cycle?

Quite by chance I noticed that my own cycle was in phase with the lunar cycle. My period was starting with the new moon. So I started looking into all the symbolic and mythological associations that have always been drawn between the menstrual and lunar cycle. I wanted to know whether they were just romantic connections or if there was more to it. This led me to involve other women in some research.

It became a project called 'Water Into Wine'. I started off by duplicating a letter in which I asked if women would be prepared to record the correlations between their menstrual cycles and the cycle of the moon.

I sent the letter to all the women I knew, and to women's centres and art colleges, where I thought there would be interest. In the end twenty-seven women wrote back, which with myself made a nice, round, lunar, twenty-eight.

What were the results?

Five were on the pill and their cycles were artificially regulated. Of the rest, none had an absolutely regular cycle. Seven had an average length of twenty-eight days, but this involved much longer and much shorter periods during the thirteen-lunar-month period. The shortest average cycle length was 23.3 days and the longest was 82.5. The average cycle length was 32.3 days which is two to three days longer than the lunar cycle.

So there was no direct correlation with the moon?

Well, I had a diary with the phases of the moon marked in it, so I had watched my cycle over a two-year period. I noticed that my own cycle was on average thirty-four days long, which is four to five days longer than the lunar cycle. Yet my period almost always occurred at the new or the full moon. So my cycle was lengthening and shortening to stay in phase with the moon. It didn't come at either of the lunar quarters as you'd think it might.

I'd only noticed this effect since living in the country, away from artificial street lighting. At least two of the women in 'Water Into Wine' showed a similar pattern, and one of these lives away from artificial street lighting.

It seems that the light of the full moon acts on the pineal gland, which is light-sensitive, and this triggers the hormones that produce ovulation. Artificial street lighting confuses this process, but in those societies without it women would respond to the moon and probably ovulate and menstruate at the same time.

It seems that close physical or emotional proximity between women also synchronises periods. Some women who took part in my project said that they had periods at the same time as other women who lived in the same house. And others noticed that if they went to stay with their mothers or sisters they often had periods at the same time as them.

Where is the pineal gland?

Behind the forehead, above and between the eyes. It's the spot that a lot of Eastern religions identify as the third eye. Perhaps they knew in some way that it was light-sensitive, and connected with insight in some way.

Is there any evidence of this function of the pineal gland?

There was a system worked out in America which involved simulating the lunar cycle in the city. You sleep in the dark in a room with light-proof curtains. But at the time of the full moon you sleep with a dim artificial light to simulate moonlight.

I've tried it. At the time I thought I was crazy and it was one of the most 'lunatic' things I'd ever done. But I was amazed to find that it worked.

These days I try and sleep in a room that faces south and gets the moonlight.

What else emerged from your project?

Well I'd also asked the women to record patterns of emotional and physical feelings at different stages of their cycle.

As people began to send the information back it became clear that this side of it was more important than any correlations with the phases of the moon.

Many of the women found that just before menstruation their consciousness underwent a marked change. I'd noticed this in myself, and I call it the 'menstrual state' because I can't think of another way of describing it.

It's a positive state and quite unmistakable. One letter described it as an unfocused elation and heightening of the senses. Colour, sound and smell are all intensified. Sexual awareness is increased. Thought seems to be experienced physically. There's no longer a separation between self and physical environment. It's a time for ideas and decision-making, and a time of increased creativity. There might be violent swings of emotion, which is greatly intensified. The menstrual state involves a different way of looking at the world. The state may be experienced as a visionary, ecstatic state. It can involve a withdrawal into the self, into a state of reverie and almost trance-like meditation.

Peter Redgrove and Penelope Shuttle point out in their book, *The Wise Wound*, that in other societies this state has been used as a source of prophecy and inspiration, and that women shamans and witchdoctors have used it as a source of power.

You can imagine how positive it was to discover that this was a common experience. When you think it just happens to you it feels like some kind of madness.

Had the people who answered your letter thought about these things much before?

Very few. Some of those who suffered from P.M.T. or period pain had thought about it. And a couple had recorded their dreams, but not in relation to their menstrual cycle.

109

I'd noticed in myself that I would wake from dreams as my period was starting. And the imagery of these dreams was very different from that which occurred in dreams at other times. The symbolism was very obvious. For example, once I dreamt that I was standing at the edge of the sea when a wave broke over me, soaking me.

I began to identify a range of images and symbols that recurred in these 'menstrual dreams'. The sea and all things associated with it predominated: lying at the edge of the sea, the tides, all sorts of water creatures. Red woman figures also.

All these dreams are much more vivid than normal ones. Colours are brighter and more emotional. And they are more easy to remember.

Now I'd always thought that these dreams were personal to me and just came from my own experience, but I was astounded to discover that other women were having 'menstrual dreams' about the same kind of things. Some were also woken from dreaming by the onset of their periods, like me.

Why does this special dreaming occur around the period?

Menstruation is a time when you're very much in touch with your unconscious. So it's not surprising that symbols from the unconscious would crop up at this time, in daydreams as well as 'sleep' dreams.

Did all the women who wrote back have this response?

A couple didn't notice any pattern at all to their dreaming, but they were on the pill, which has a drastic effect on the menstrual cycle, virtually suppressing it.

Don't you think that this glowing view of menstruation must sound rather hollow to women who have severe P.M.T. and bad period pains?

None of the women who took part in 'Water Into Wine' escaped period pain entirely. But those who suffered least were those who felt happy about menstruation and were in pleasant and sympathetic surroundings during their periods.

We live in a society in which menstruation is taboo: it's negated and we generally pretend that it doesn't happen. We have a process going on inside us which we've been made to feel is unpleasant and unclean. Now if you're doing something you can't help, yet society tells you it's horrible, this has a very strong emotional and physical effect on you. It makes you feel bad.

Women internalise this fear and disgust and I think this is a major cause of P.M.T. and period pain. P.M.T. and period pain increase women's self-disgust, which in turn increases P.M.T. and period pain.

I think P.M.T. and period pain are the opposite, or a strange distortion, of the menstrual state which has been blocked by the menstrual taboo.

Surely difficult periods don't just come down to the nature of the taboo? There must be physiological causes as well.

I'm not saying that there aren't. This is obviously the case for some women. But equally, if you're encouraged to pretend that something that's physically, mentally and emotionally happening to you isn't happening to you, then it will turn back on you and become painful and difficult.

P.M.T. and period pain are also to do with lack of time and space. If you're involved in an arduous work schedule, or you're at home on your own with three young children, you have no quiet time in which to relax and withdraw into yourself to dream or contemplate.

'Water Into Wine' showed that if we did have this time and space menstruation became positive, but if we had constantly to be thinking and worrying about other things, then tension and pain were often the result.

We all now try to make that space for ourselves, but for a lot of women this is just not possible. Our society should make this space.

In many other cultures women actually had a place, sometimes called a menstrual hut, where they went to be on their own during their period, away from the pressures of childcare and other work.

The first accounts of these were written by male anthropologists. They'd found these huts among American Indians and in New Guinea, and assumed that women were segregated in these places during menstruation by men who felt they were unclean and didn't want anything to do with them. In some societies this was so, but subsequent research has shown that in far more cases it was the women themselves who wanted to be on their own.

Isn't there a danger of the 'special case' argument here? Women have been struggling for equal pay and conditions for the past twenty years on the basis of their equality to men. To demand special conditions for women seems to play right into men's hands, giving them a rationale to sustain inequality.

But the struggle for equal pay and conditions has been on men's terms, hasn't it? Work and time in this society are organised without reference to women's cycles. I'm interested in reorganising working patterns so that menstruation is seen as a natural part of the time cycles of half the population.

Am I supposed to act differently now?

At the beginning of the Women's Movement, the last thing many women wanted to admit was that menstruation made them any different from men. They wanted to pretend it didn't happen and that they could do anything men could do. Now I think that a lot of women are coming to realise that they can do things that men can't do, that menstruation is important, it's theirs, and society must make room for it. At the moment time is organised in a male way.

How would you characterise male time?

It seems to be more about linear progression. It has little reference to natural or biological cycles. It's hard to know whether this is innate or whether it's something that's developed over the past few centuries.

What do you feel about men walking on the moon?

It's just the sort of thing they would do really, I suppose. What's the point of walking on the moon, when most men don't even know how it affects them and the women they live with.

DAILY EXPRESS

MONDAY JULY 21 1969

Weather: Sunny spells; very warm

No. 21,499

Price 4d.

2am: 'We'll walk now'

MAN IS ON THE MOON

NEIL ARMSTRONG   EDWIN ALDRIN

TWO hours after landing on the moon last night Neil Armstrong and Edwin Aldrin decided that they would step on to the lunar surface at 2 o'clock this morning.

DATELINE: Sea of

POCKET CARTOON

LATEST

The complete story

Officials at Houston gave the time taken on the flight from earth to the moon as 102 hours, 45 minutes and 42 seconds. With the landing confirmed, a controller said : "We're breathing again. But we've got a bunch of guys here at Houston about to turn blue."

Ground Control said the landing took place 81 seconds earlier than scheduled. They told the two men that all systems of their craft "look pretty good" after the touch-down.

But there were two problems. Some fuel had been trapped in the pipe line feeding the descent rocket engine. Ground Control "did not think it critical." And their clock had jammed. It was repaired later.

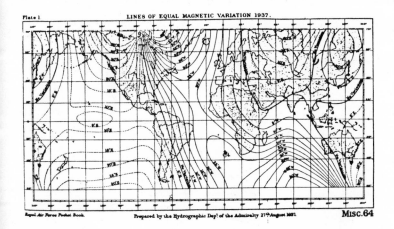

The Western mind is hooked on diagnosis: to understand the cause of an effect is deeply reassuring. It reaffirms the belief that everything can be known, and therefore, the hope that we're in control – a delusion that characterises Western science.

The first reaction of rational minds to the idea that the moon influences human beings is to seek causal explanations. In fact, it's quite easy to come up with plausible scientific hypotheses. With solar minds we can 'bring light to bear' and colonise the lunar mystery.

So: moonlight affects the timing of menstruation because of the action of photons on the pituitary gland. (And why not? Sunlight causes plants to grow through photosynthesis.)

Or: the fluctuations in the gravitational pull of the moon, due to its cycle, have a direct physical effect on patterns of dreaming and the menstrual cycle. (And why not? The moon causes the tides, and lifts the sands of deserts in the same way.)

Or: the moon's orbit causes variations in the earth's magnetic field, which in turn has a cyclical effect on the electrical functioning of the human brain and nervous system. (And why not? Compasses point north . . .)

These thoughts 'make sense'. We can stamp them 'LOGICAL', and place them confidently in the in-trays of ultimately testable speculations.

But rationalism can't patronise the three conversations that follow, which focus, among other things, on witchcraft, astrology and the I Ching. For these three different supernatural traditions of knowledge are mutually concerned with reading the possibilities of the present and the future in a way that rejects the omnipotence of cause and effect as we know it, and embraces views of time that challenge the authority, in our culture, of sequential linear time.

As we'll see later, twentieth-century developments in physics and recent hypotheses in biology affirm the 'irrational' view of time and causality that's at the heart of these ancient practices. But for the time being we need look no further than the world of our unremembered dreams, where we keep the irrational and a-causal centre of our emotional lives well closeted, out of the daylight.

In the three conversations that follow, we talked to Marian Green again – who is in fact a witch; Liz Greene, who is an astrologer; and Marie-Louise von Franz, who is a Jungian analyst. All three are writers and all, in their different ways, are therapists. In another culture, they might have been called 'wise women'.

What does the moon mean to a witch?

The new moon is the time to begin
things. This could be looking for a new
job, or planting seeds.

The full moon is the time of completion.
Many of the old witch ceremonies were held
on the night of the full moon because it was a
time when it was safe to travel. You could see
where you were going. It was also an easily
recognised date: anyone knows a full moon even
if they're less familiar with the other phases.

I think anyone who observes the way they dream will
find that around the time of the full moon their dreams
are more vivid and easy to remember.

With the waning moon you might begin a healing, so that
as the light of the moon disappeared, the fever would go
with it.

Witches also knew about the connection between the phases
of the moon and a woman's menstrual cycle. She could often
advise a woman when the best time was to conceive.

What is your function as a witch in the modern world?

We try to preserve the ancient arts of healing, divination, keeping up the old festivals and looking into the future. At the centre of this is an attitude to working with nature. The old witches believed that the earth was a living goddess and mother, and that if you lived on the surface of your mother you had to be tender and careful towards her. We tend to forget this, which puts our future in the balance. We tend to think that we're living on a lump of rock that's floating through space. We need to get back to working in harmony with nature.

This all means that our work is closely connected to the seasons, the changes in nature in the turning year.

Witchcraft also tries to teach people to balance the inner, darker side of their nature (the one that's threatening the planet!) with their outer, brighter side. Most people are aware that they have an untamed, wild side. If they ignore it and allow it to be repressed, it can override the brighter, outward side of their nature. But if they learn to come to terms with this force, it can become a great source of energy.

How did you decide to be a witch?

That's a hard question to answer. It's something that creeps up on you, like political conviction or religious belief. But I know that at school I was looked on as odd, and my classmates used to call me a witch. I learnt my first bits of magic as a teenager, and later on I was trained by witches. I've worked with a coven and been through formal ceremonies. But I see myself more in the tradition of the

We see the moon as the symbol of the goddess. And she is changeable, having three jobs, as it were. At the time of the new moon, with its narrow waxing crescent, she is hope and promise, the beautiful young girl who hasn't been touched. At the full moon she is the pregnant mother who we all interact with. The waning moon is the old crone who is lonely. She is very wise, but she isn't easy to get on with because she answers your questions truthfully.

village witch. I think being a witch is more an attitude of mind.

Who was the village witch?

She wasn't only a herbalist. She also had the almost ritualistic functions of midwife and layer out of the dead. And she knew how to prevent conception or cause abortion. In a pre-literate society, in which knowledge was only held in the memory, these powers of life and death were very special and go some way towards explaining why people feared witches.

The village witches were often psychic and had the power to read the future. To many this appeared to be the power to control the future, which also explains why they were feared and persecuted.

Where does the word 'witch' come from?

The three main theories are that it comes from either 'wit', meaning wisdom or wiseness, as in wizard; or from the Anglo-Saxon world 'wicca', for a person who had ancient knowledge; or from the Old English word 'witcher' which meant crooked or bent, one who has departed from the straight path.

In French, the midwife is called the *'sage femme'* meaning the wise woman.

Which definition do you prefer?

I'll settle for all three.

How do witches see into the future?

Mainly by training ourselves in auto-hypnosis or deep meditation, techniques by which you use imagery to free your spirit from your body, so that it can travel through space and time.

Forward through time?

Sometimes. Yes.

To something definite, or just a possibility?

No one can know exactly how the pattern of the future is going to work out. Life is a bit like a chess game played in four dimensions: every time someone makes a move (and there are more than two sides in this game) the whole pattern is altered to show different possible futures. A witch learns to read which trends in the possible futures are likely to remain unchanged.

How do the techniques of auto-hypnosis differ from Eastern forms of meditation?

They're probably not very different at all. It's just a question of how you alter your state. In the East, people use the chanting of mantras, or controlled breathing. In the West we have tended to use imagery: you imagine a journey, or meeting a wise person or guide.

Do witches use hallucinogenic drugs to achieve these states?

We tend to regard them as short cuts because you are not in full control of yourself when you take them. If you're high for twenty-four hours people might have to wait rather too long for a coherent answer to their question.

The old witches used to make a potion called flying ointment. This was a mixture of fat and the juices of various herbs which would be rubbed on the hands and body until they became numb. So you wouldn't be able to tell whether you were sitting or standing, and you almost had the sensation of flying. In fact it gave you visions of flying, hence the witch's broomstick. And when you went to bed you dreamt vividly.

Witches would also use various kinds of hallucinogenic mushrooms which are to be found in the woods around Hallowe'en. The *Amanita muscaria* has long been recognised by people all over the world as being a mushroom that increases the ability to dream and have visions. But you have to know what you're doing with it, otherwise it can be very poisonous. The shamans of Russia used it so that they could bring messages from the gods to the tribe that they were serving.

How do these various mushrooms and potions affect the perception of time?

They dilate time, like a dream does. They enable you to step outside time as do many of the traditional methods of clairvoyance.

What is the witches' circle?

Witches and magicians have always created circles. It is a space swept clear with the witches' broom, and marked out with stones or pieces of wood. It is a special space like a surgeon's operating theatre, free from outside interference. If you make a circle, and you cleanse and seal it symbolically, you create a spot where subtle psychic influences don't get in the way. There won't be stray thoughts or atmosphere left over from what has previously happened in that space.

You can make a circle anywhere, but the wilder, older places are better, where things haven't been played about with too much by people, and it's easier to get a quiet psychic atmosphere.

Once you enter the circle you're in a different time zone, and you're dealing with spirits or forces that don't exist in our everyday dimension. You're set free from clock time to meet with other beings in their time. When you're inside the circle normal time stops. The old stone circles were probably magic in this way.

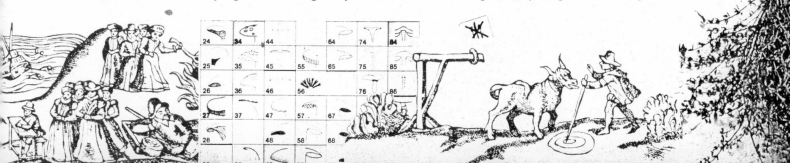

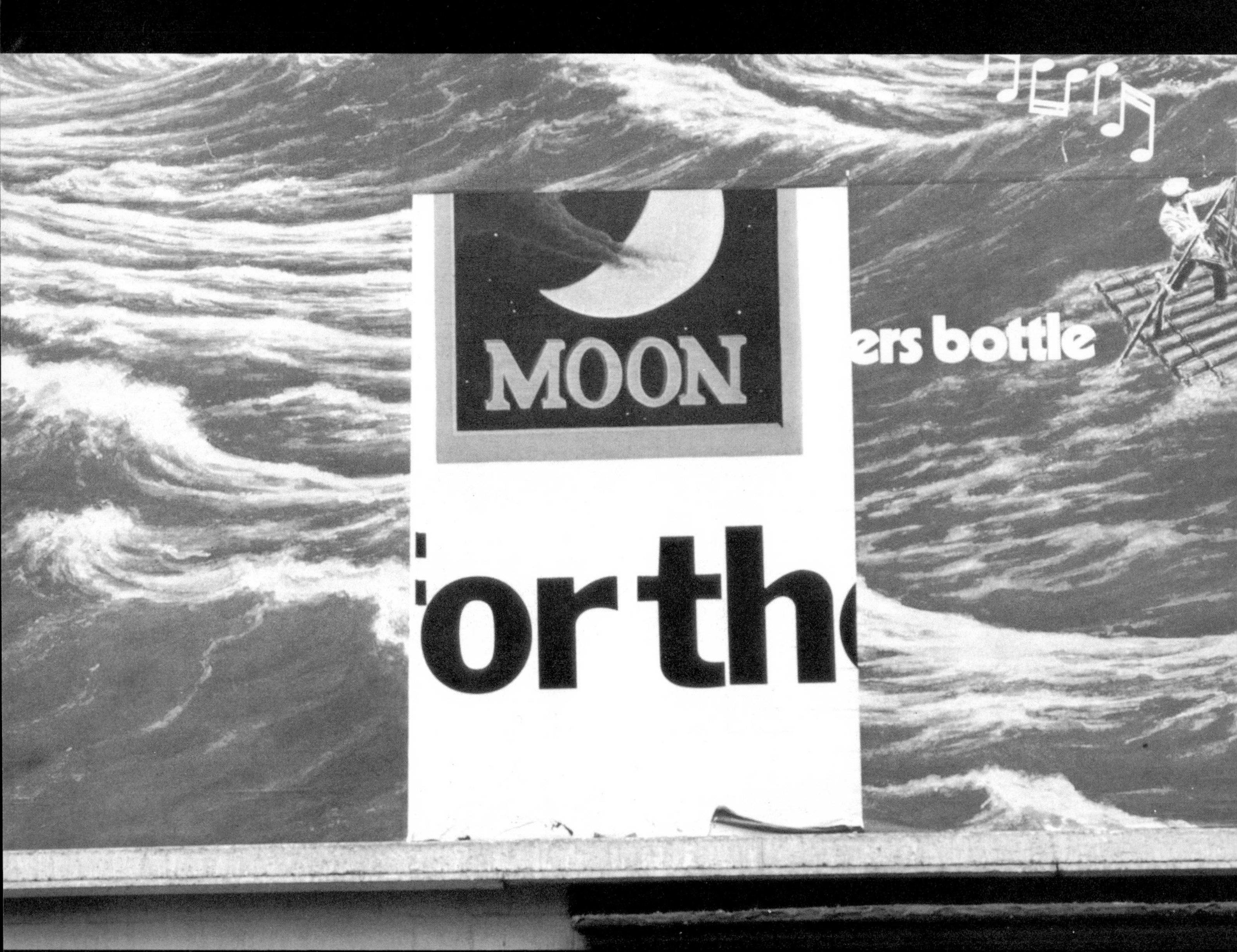
MOON
ers bottle
for th

'Careful! It's a new moon,' my mother would say, and out of the house we'd all troop. Or we'd just draw the curtains. She was convinced it was bad luck to see the new moon through glass, but in fact, like everyone else, she had the superstition wrong. It's bad luck to see the new moon in a mirror – to mistake the waxing moon for a waning moon.

There were so many little magical codes of practice. A single magpie was bad luck, but a spit cancelled the danger. A piebald horse sanctioned a wish, which wouldn't be granted if you saw its tail. To pass on the stairs was disastrous, so the one going down would wait at the top for the one coming up. Pinches of spilt salt thrown over the left shoulder and evasive action round ladders were second nature. And if, by chance, two of us spoke the same word simultaneously, we had to shake little fingers silently to undo the harm that would follow. Wood, of course, was endlessly touched to insure against undue optimism: plastic simulated woodgrain wouldn't do.

All these 'old wives' tales' went with a kind of sheepish light-heartedness. Superstition was given the veneer of a joke, as if in the twentieth century such practices were slightly mad. But they were very serious – the fun was there to make light of taking such irrational things so seriously.

Today I still spit at single magpies and wish on piebald horses. I've added a few obsessions of my own as well: it's bad luck if I don't reach the other side of the road before the pedestrian lights stop flashing. And I avoid the new moon in a mirror.

I'm not on my own with these quirks. In the heart of our allegedly rational culture these little domestic acts of magic are rife. In the midst of all our gleaming technology, which appears to enable us to shape our destiny, we're still fighting evil spirits. Why does a culture that champions the value of rational science resort to the irrational to influence future events? Could it be that we've deluded ourselves as to quite how rational we are?

The immense and growing popularity of astrology reflects this irony. There are more birth charts drawn up today than ever before – there are even instant computerised charts available from booths on railway stations.

The popular assumption about astrology is that it is an irrational way of predicting events rather than influencing them: such and such will happen because it has been predetermined by the exact positions of the sun, the moon and the planets in relation to your precise position on earth at the moment of your birth. In other words, our lives are predestined; there is no such thing as free-will. If you know your chart, you know your future.

Liz Greene took up these misconceptions about astrology in our conversation with her. We began, once again, on the moon, but ended up in the realm of the psyche.

LIZ GREENE holds a Ph.D. in Psychology and a Diploma in Analytical Psychology from the Association of Jungian Analysts in London. She works as both a professional astrologer and a Jungian analyst, as well as being a co-director of the Centre for Psychological Astrology which offers a training in astrological counselling. She is the author of several books on astrology and has also published two historical books.

What does the moon mean for you?

In mythology it is very mysterious. It seems to be associated with hidden things: the night, the unconscious, the dream and fantasy world. It represents a psychological dimension other than that of the light of day.

Does it affect you in any way?

It's easy to be accused of twisting facts to fit the imagination with this kind of question. But I can usually tell when the moon is going to be full because, even if I haven't checked to see, I tend to get into awful moods. The full moon tends to heighten my energy and makes me a lot more active. But it also makes me more sensitive and impatient. I become more aware of the cyclical nature of my moods.

Emotional build-ups and arguments often happen around the time of the full moon. It seems to create a certain amount of jamming in the works. Trains get delayed and accidents happen. It's been known for a long time that psychiatric wards get very jumpy when the moon is full. I'm also aware of these mood changes in the people whom I see therapeutically. Things tend to be a lot calmer when the moon is new.

Why do you think that the moon is seen as feminine in so many cultures?

It has an evocative quality on the imaginal level. It is mysterious, it appears at night, it has many faces through its phases, it disappears only to appear again. This waxing and waning give it associations with fertility, pregnancy, and birth.

In myth we find that many goddesses are lunar goddesses, and they are often portrayed as pregnant, or they are goddesses of childbirth.

The goddess who was worshipped as the moon was also the protectress of the fertility of the earth. In fact the moon seems to have had deep connections across the Mediterranean with agricultural societies in which crops were planted and harvested according to lunar cycles.

The same is true of prehistoric Britain, where Harvest Festivals were always linked to the lunar cycles.

Easter, with its motifs of eggs and resurrection, is associated with fertility, and is a lunar festival. It still always falls on the Sunday after the full moon when the sun has entered Aries. This is close to the vernal equinox, with all its associations of spring and the emergence of new crops. In farmers' almanacs we find the suggestion that seeds should be sown when the moon is waxing rather than waning. And there has always been lore about harvesting herbs when the moon is in the right phase.

I am quite prepared to believe that these things are valid. After all, the moon certainly has a physical effect on the tides, and also on the human body. My experience of keeping pets confirms this; whenever my gerbils were pregnant, they would always give birth at the full moon, like clockwork.

123

If the moon is feminine, does this mean that it has nothing to do with men?

I think it would be a great mistake to take the word feminine in this way. The feminine is a principle which is at work in the whole of life. It will have equal importance for plants, societies, people and cities – for anything that lives. I don't think it's right to draw distinctions in that way. The feminine principle is very much alive and well in men. But if you are very bound up in seeing men and women as so radically different that there's no point of connection between them at all, then the feminine is projected wholly on to women. But I think a man's moods are just as changeable, just as much connected to a cyclical psychic movement, as a woman's. It's just that this isn't talked about, because culturally it's not acceptable.

No, my interpretations of the moon in astrology are that it is concerned with the feminine side of an individual's nature, whether the individual is a man or a woman.

How did you get involved with astrology?

Someone took me to an astrologer. I think that's the way most of us get bitten. I was studying psychology at university at the time. The emphasis was on behavioural psychology and I was learning to observe people's behaviour from an environmental point of view. When I had my chart done I was very perturbed because the astrologer got information about me from the chart which she could not possibly have obtained through any rational means. This baffled me and I wanted to find out more about it. What became apparent to me was that astrology offered a far better psychological map than behavioural psychology could. I found it much more profound, allowing much more dignity to inherent individuality, yet without negating the shaping influence of the environment.

Has astrology affected the way you think about time?

It has made me look much more carefully at the cyclical nature of things, because astrology is about cycles.

Looking at birth charts, you discover the tendency of things to come back again. Instead of seeing experience A that I'd been through and experience B that I was going through as having nothing to do with each other, I could see that the planetary positions for both events in the chart were similar. There was a very definite connection between A that had happened in the past and B that was happening in the present. Apparently diverse events can have the same meaning. So astrology has made me see my own life as much more interconnected. And it's taught me to look at other people's lives this way as well.

How do you do someone's chart?

The drawing up of a chart is a mathematical procedure based on astronomical reality. To do it properly, I have to know the birth time, place and date as precisely as possible. Once I have these, I consult an ephemeris, which is a kind of planetary timetable, to find out the precise positions of sun, moon and planets at this moment in space and time. The ephemeris is set for Greenwich midnight every twenty-four hours for every day of the year. It's a very precise measurement; wherever in the world a person is born, I have to convert the time back to Greenwich. Then I can locate where each planet was for any birth chart I might want to do.

Why is the precise time so important?

The birth chart freezes the eternal movements of the planets at the particular moment when the individual begins life. This is usually associated with when the person takes the first breath; that is the beginning of life as a separate entity. This is the moment which is photographed, as it were, and the chart simply maps out where everything was in the solar system as seen from the place of birth.

A popular assumption is that this 'snapshot' indicates the predetermined way in which everything in the individual's life will fall out. How do you reconcile free-will with the fated view of life that astrology seems to offer?

I think this is the source of the worst misunderstanding

about astrology on the part of the layman. The placements in the chart don't reflect a series of events that are going to happen. They describe a psychological dynamic: an image of the motivations, cross-currents, drives, conflicts, needs and attributes within an individual. As Jung so nicely put it, a man's life is characteristic of himself. If one is made a certain way, one is going to seek certain kinds of experiences, according to one's nature.

So the chart doesn't map out events that are going to happen. It maps out what the person is like, and the phases of development. There's nothing particularly supernatural about this. We take it for granted in gardening when we plant a seed. We have a rough idea of when it's going to sprout, bear leaves, bear fruit, of what it's going to look like. Basically, a chart works in the same way. It maps out the development of the organism on an inner level. There are psychological changes that accompany biological changes, at puberty and menopause for example, and there are psychological changes that don't have physiological reflections and which are unique to one individual or another. Both are natural, and the chart will describe when these changes are likely to happen.

So you would talk about an individual's predisposition to things rather than that certain things will happen?

Yes. A predisposition to respond to life in certain ways, to wind up in certain situations, to develop certain kinds of attitudes. And in this sense, far from being fated, one is very responsible for one's life. The only thing that fates us is ourselves.

How much credibility would you give to what you saw in my chart as against the social conditioning of my family and culture, the economic circumstances of my life, and so on?

All that would have to be taken into account. It's back to the old question of inherent temperament versus environment. I think that both are equally important. A particular chart is going to have a chance of

ZORITA

The famous night club entertainer does her "Half and Half" dance.

unfolding better in some environments than in others. And the chart won't tell me whether you are a man or a woman, and that's obviously going to make a big difference as to how you use the potentials of the chart. Nor will it tell me your racial background, and that will be an additional factor to look at. The art is really to put all these factors together.

That's why I don't like playing guessing games when people want their charts done. Someone might say to me, 'I'm not going to give you any information other than my birth time, place and date, so that you can prove to me that it works.' But I think it's much more fruitful if I know something of the person's setting in life. Then I can conjoin that with the chart and offer genuine help, instead of a parlour game.

Would you be likely to look at a person's chart and tell them that in ten years' time, say, they faced a period of illness?

I wouldn't know that it was going to be an illness. But I might see a period of crisis, where certain kinds of conflict come to a head, and if the person was not able to handle this, then illness might be one outcome. Illness is one way of working things out. There are other ways of working out the same conflict, and you can't really tell from a chart which of those ways it's going to be, because a person always has the option to start examining what's going on. As soon as one begins to tamper, it becomes very difficult to predict a concrete thing like illness.

So astrology is more about a choice of probabilities rather than predicting the certain outcome?

Yes. As in quantum theory, the observer affects the experiment. The moment one started contemplating this

crisis in ten years' time, one starts thinking about where one is going and who one is. One has already affected the outcome.

A lot of people get very agitated with astrology because they think it means that we can't influence our futures. But you're suggesting that we do have free-will: that because I and my future are observer and observed, as it were, I will influence future probabilities in my life in the act of working on them in the present.

It isn't 'influence'. Being conscious of something that affects you in turn affects both you and that something. If I spend a bit of time with you and try to understand who you are, it's going to affect both of us in some way, and therefore it will affect the outcome of our encounter. It's not so much 'influence', which is about trying to control things, as about co-operation. It's more like learning to dance – with oneself.

How do you explain the apparent connection between the macrocosm of the planets and the microcosm of the individual?

I couldn't possibly explain it. Nor, I think, could any other astrologer. But the connection appears to exist.

Future Forecasts

Personal Horoscopes

VOX STELLARUM:
OR, A Loyal
M A N A C K
For the Year of Human Redemption
M.DCC.XCI.

She was his wife and he loved her. But if she mentioned fate, or destiny, or fortune-tellers once more – so help him, he'd scream . . .

CANCER (June 22 . . .

How would you describe the experience of time in dreams?

Dreams appear not to pay attention to what we understand as time. Perhaps that's why they are traditionally associated with the moon and the imagination. A dream will jumble past, present and future as if they were all happening at once.

This kind of 'timelessness' is not only experienced in dreams. It can happen when one is doing something creative, or when one is very engrossed in something, or when a person goes into a fantasy or a deep recollection. I see it a lot in my analytic work, where experiences that took ten years to live through are re-experienced in thirty seconds; and all the feelings are happening right there in the present moment.

This suggests that the past is somehow locked up in the present.

Yes. This becomes evident when I work deeply with someone on feelings which have been repressed, pushed into the unconscious, so that they are not accessible to rational waking consciousness. When these feelings rise to the surface, they are happening now. When people start remembering childhood, they are not simply recalling something that happened in the past. They are experiencing past feelings in the present and one can feel the young child sitting there in the adult body. It's a kind of simultaneous experiencing of the whole of one's life, now. This is why I think it's a mistake to look at psychological work as digging up the past, because it isn't the past.

I think feelings tend to exist in this timeless frame. And they come out in present-day relationships even if they belong to the past. Quite inappropriately sometimes, anxieties or expectations will arise that have nothing to do with the person in the present moment. The past is being lived, but it's happening now.

But fear and anxiety are time-related emotions, aren't they?

There is a kind of anxiety like 'I'm going to miss my plane', and this is time-related. But the sort of anxiety that deeply disturbs people and brings them to seek help often comes from a timeless place which is not connected to the ordinary time-frame which we think of.

How would you describe that 'ordinary time-frame'?

I understand it as you arriving at 10.00 a.m. this morning, starting filming and finishing by 1.00 p.m., and me having an appointment at 1.15 p.m. And when I wake up tomorrow, today will have become past. It's a sense of events passing back into a place where they are no longer alive, no longer part of the present reality. That's the way I would understand the passage of time in the ordinary sense.

Whereas the time of the unconscious is jumbled like dreams?

Yes. I find this hard to grasp, because like everyone else I see time as linear. What I ate for lunch yesterday was yesterday. What I eat for breakfast tomorrow will be tomorrow. But the order of sequence in dreams often seems back to front.

Now, this can give rise to a feeling that things are fated. If we can see it in a dream, then somehow it must be already planned in the boardroom and we have no choice in the matter. But I don't think this is really what is going on. It seems that the unconscious doesn't have the same time sense as the ego. It's more of a timeless place where everything is occurring at once.

What do you mean by 'ego' and 'unconscious'?

I use the word 'ego' to describe my sense of me, my consciousness, my awareness of myself. And if the ego is what I am conscious of as myself, the unconscious is going to describe all the other aspects of my psyche of which I'm not aware but which are nevertheless part of me. They may be potentials, feelings, drives, appetites, convictions, anything you can think of that belongs to me but which are, as it were, behind my back, so that I can't see them.

If the unconscious, through our dreams, can foretell the future, what about free-will?

That is the problem. Working as an astrologer, and having some familiarity with the so-called divinatory arts like the tarot and the I Ching, it looks to me as though one can foretell the future. But I suspect this is not really what it is about.

In the timelessness of the unconscious, the future seems to happen in the present, because its seeds are there. The future is the natural flowering of the past and the present. There is a level at which an acorn, for example, contains all the potential cycles of development of the oak tree. As an astrologer, I would understand an individual to contain his future potential in the same way.

What did Jung mean by the 'collective unconscious'?

I think he was taking the idea of the unconscious further, and postulating that there is a level at which human beings share a psychic unity. Like at a football match, for example, where one gets the feeling of everybody being joined up as one by certain emotions.

Central detail from Diego Rivera's mural in the Hotel del Prado in Mexico City, 'Dream of a Sunday Afternoon at Alameda Park'. The skeleton's right hand is held by Rivera himself as a boy. Directly behind him, with a hand on his shoulder, is seen – as an adult – his wife, the painter Frida Kahlo. (Next to her is the Cuban patriot, José Marti.)

A kind of overlapping of minds?

It's not 'minds'. Probably it's more a connectedness of emotion and instinct between people at a very deep level. For example, one culture will produce a certain kind of religious or mythological image, and another culture, 9,000 miles away, will come up with roughly the same kind of image at roughly the same time, when there has been no communication between them.

For example?

From South America to Greece, Babylon, China, Japan, we find figures of lunar goddesses, and they are always connected with dogs, the metal silver, the moon, childbirth and witchcraft. There is a similarity of motif that one cannot justify by cross-cultural migration.

We also find this in fairy tales. The same stories crop up in American Indian folk tales, Japanese fairy stories, and the Grimm Brothers. Of course the names are different, the stories have their particular cultural overlay, but they are the same. They are stories which have come up spontaneously, that have been told at firesides over many generations. These seem to come from a level of the human imagination that is common to all cultures.

una   lunacy   luna moth   lunar   lunate   BLOOD MOON

moonshine   moonstone   moonstruck   HUNTER'S MOON

mo

R'S MOON

OLD M

WANING

OON   SA

ON   MO

TER MOON

FUL

HONEY M

n

d

OON

RESCEN

BLUE

E DEER

GRASS

OD

W

moo

moonie

moonstri

y   moo

moont

ch   mo

D   MOO

ON   ROSE MOON   MOON WHEN T

lunatic   HARVEST MOON

RS   BIG WINTER MOON   GREEN

We all treasure our tales of fantastic coincidences. This one is true.

I had recently broken off a relationship with P. which deep inside I had wanted to continue. I was miserable. As the weeks dragged by after the separation, I clung on emotionally, nurturing past letters and photographs. I lived in London and she lived in Bristol. I'd made an awful mistake.

One day I was walking down Tottenham Court Road sharing my unhappiness about P. with a friend. I was telling her a dream that I'd had the night before. It went like this.

I was wandering with P. in the rain on a farm. Leaving the buildings behind, we set out through thick mud towards a knoll that lay close by, at the foot of a range of low mountains. It was very wet and low cloud clung to the landscape. We squelched up the hill in grey plastic macs.

At the top of the knoll was a long, zinc drinking trough for cows. When we reached it we looked down into the water and saw a hairless pink piglet submerged. It should have been drowned, but its eyes were open, looking up at us, apparently alive.

I'd just finished recounting this dream to my friend, half-way down Charing Cross Road, when I looked up and saw P. flash across our path and disappear into the doorway of a bookshop.

I froze, and then started shaking. My friend turned to me and asked what was wrong. I said, 'That's her. P. She's just gone into that bookshop. Please. Please will you go inside and see if she's in there.'

I leant against a lamp-post as my friend went in, convinced I was hallucinating. But two minutes later she emerged with P. It was her.

The trouble was that I had no place in my rational outlook for such an event. I was somehow ashamed to embrace the coincidence. To my way of thinking, cause and effect ruled the roost, and anything that didn't fit the logical scheme of things I denounced as 'unscientific', 'idealist' or 'undialectical'.

In reality, though, I had little choice. The experience was already inside me.

The relationship restarted some months later and the coincidence became one of those dinner-table stories that compete with each other for 'far-out-ness'.

In our conversation with Marie-Louise von Franz, we started talking about synchronicity, which is Jung's word for the kind of coincidence I've described. It led her on to her lifelong involvement with the I Ching, and a wide-ranging discussion about time and the limitations of Western causal thinking.

MARIE-LOUISE VON FRANZ is a practising psychoanalyst and lives in Switzerland. For many years she worked with C. G. Jung. She has written books on Time, Fairy Tales, Dreams and Alchemy. The I Ching has had a central presence in all her work. She is now completing a book about Death Dreams.

MARIE-LOUISE VON FRANZ – SYNCHRONICITY AND THE I CHING

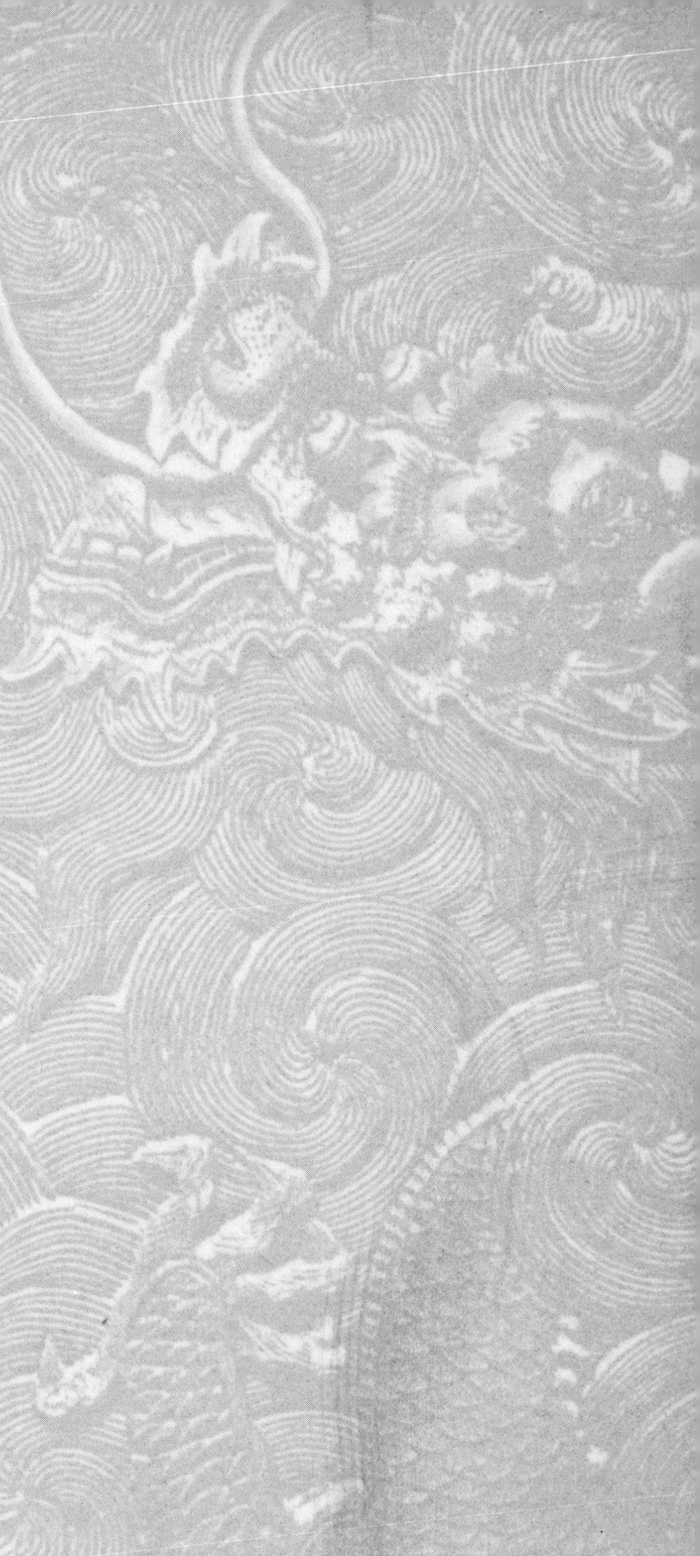

What is synchronicity?

Synchronicity is a term Jung invented to describe a meaningful coincidence. Many coincidences have no meaning: for example, if I blow my nose at an airport and a plane crashes. But if I had a horrible foreboding that a plane was going to crash and at that moment it did, that would be a meaningful coincidence. My foreboding wasn't caused by the crash, nor was the crash caused by my foreboding. There was no causal connection, yet the coincidence is meaningful.

'Meaningful' suggests that there are connections beyond the appearance of things that are non-causal. This is quite hard for the Western rational mind to accept.

Yes, because we are causally trained. We think that B comes from A, C from B, D from C . . . and so on. I was once giving a lecture on synchronicity and there was a Japanese professor in the audience. At the end he thanked me and said: 'for we Easterners synchronicity is self-evident and we don't need a lecture on it. But by explaining the difference between synchronicity and causality, I've at last understood causality!'

You see, causal thinking is simply a habit. If you're walking along a beach and you find an old hat, two shells, a bit of wood, and an empty bottle, you might seek a causal explanation of how they got there. But the Chinese would ask: 'Why have I arrived when all these things are together? What does it mean?'

In ancient China the I Ching was a way of answering such questions. It is based on the philosophy of Taoism. To be 'in Tao' means to be in harmony with the moment in time. The I Ching is a technique for inquiring into the quality of the moment, and it adds a philosophy of what to do.

I learnt the I Ching when I was nineteen and it accompanies me wherever I go. It's my Bible!

How do you consult it?

With a question in mind I throw three coins. They must really roll so that chance can have its chance. I get so many heads and tails. The tails are worth two and the heads three. By adding them up and throwing six times, I arrive at a hexagram. By looking up the hexagram in the I Ching I can read, so to speak, the quality of the moment.

There are moments when I should lie low, moments when I should assert myself, moments when I should be modest, or fight, or give in. The whole idea is that no action is good or bad in itself; it all depends on the moment in which it happens.

AS WE FILMED THIS CONVERSATION, MARIE-LOUISE CONSULTED THE I CHING.

Have you just asked a question?

I didn't 'ask', but I had a question in mind. 'What situation are we in here?'

Did you get a helpful answer?

It's too intimate to tell you, but it's an answer for something I had in mind.

Was the answer positive about our situation?

It was Number Thirty-four 'The Power of the Great'. It's difficult for me to talk about, but I asked what it means for me that more and more T.V. companies, publishers and so on, are asking me to speak. This is the power of the situation I'm in.

Something in me feels that this is dangerous, that too much of it can pervert one's character. So for me the answer is: 'You have power, but don't misuse it. Be very careful.' But it's not my power, it's the power of the situation. It's as if the I Ching is saying: 'You are in a situation where there is great power about. Tread carefully and don't get carried away.' I would add: 'And my unconscious knows it.'

What are the coins responding to when they fall? Your feeling?

That's a causal question! They're not responding to anything. They are falling by sheer chance and the quality of the moment is in how they fall.

But if you didn't feel your question, perhaps they would fall differently?

No. I think the question is already there in my unconscious. That's where the idea of what to ask comes from.

Do you throw the I Ching every day?

I once did for a while. The question I asked was: 'Please give me the hexagram that is closest to the meaning of my dreams.' I wanted to see if the I Ching and my dreams had a connection. I did this for several years and the dream and the hexagram always matched.

Isn't there a danger of becoming too dependent on it?

Certainly, but the I Ching has a good method of self-protection. Once when I was young I used it too much and I got the answer 'Youthful Folly'. That's as if to say, 'If you ask once I'll give you an answer, but if you ask twice, I won't!'

So I generally keep the I Ching for important questions.

You can also misuse the I Ching by misunderstanding it. The text is rather vague and open to twisting in the way that you want to understand it. But humans misuse everything, from computers to atomic physics.

Have you ever been involved with astrology?

No, I didn't go deeply into it because of my personal taste. But it was Jung who convinced me that it wasn't all humbug, because one day he told me where my moon was. At that time I hadn't had a horoscope made, so I had one done and Jung had guessed right. I thought that if you can guess from the person to the horoscope that there must be something in it.

In fact the basic idea of Chinese astrology is similar to the I Ching. Both are built on the question: 'What likes to coincide?' In astrology it is the coincidence of the moment of my birth in relation to the positions of the planets that is meaningful. But again, there is no causal connection. The stars don't influence us and we don't influence the stars.

You mentioned misuse of physics, presumably meaning 'the bomb'. But modern physics also seems to be confirming the non-causal synchronous view of the universe on which the I Ching and astrology are based.

Yes. There's a good example of this in what's called the E.P.R.* Paradox. If two particles which have been connected are moving apart in opposite directions, and the spin of one is changed when it reaches, say, New York, then the

The changes in spin coincide even though there is no causal connection, no communication between them.

This has recently been proved experimentally and is turning out to be a great headache for physicists. How are non-causal connections to be built into the physicists' causal view of the world? Well, we're beginning to solve this problem by saying that the universe is a unity; that what happens in any one place in the universe has a simultaneous reverberation everywhere else.

This is the Chinese view of the universe: the idea of synchronicity in which the universe is a whole and everything that happens within one moment belongs to the whole in some way. The only question is, can we read the text of these moments?

How do you explain the apparent convergence of ideas in modern physics and ancient Eastern mysticism?

I'm tempted to say that both are uncovering the same truth. I would have liked to say *the* truth but perhaps that's trying to say too much. One way of looking at it is to see the Eastern mystics as introverts, discovering this truth by looking inwards, into their own unconscious; and Western physicists as extroverts, who arrive at the same ideas by looking outwards at what happens in the cosmos.

What is this 'truth'?

First of all, that the universe is one great unity. Second, that the process of this whole universe is an energy dance. And third, that everything is an energy phenomenon. These were the intuitions of the Eastern mystics, and they also seem to be the resulting conclusions from investigations in modern physics.

Einstein, Podolsky and Rosen, see p. 168

see p. 168

135

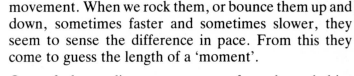

MARIE-LOUISE VON FRANZ – CHILDHOOD, OLD AGE AND DREAM TIME

How do you think children develop their first ideas of time?

When you're a baby you seem to live, feeling wise, in a timeless world, which is only interrupted by the rhythm of waking, feeding and sleeping.

Perhaps the first sense of time babies experience is a feeling for the frequency and rhythm of physical

movement. When we rock them, or bounce them up and down, sometimes faster and sometimes slower, they seem to sense the difference in pace. From this they come to guess the length of a 'moment'.

One of the earliest movements of newborn babies involves shaking the arms with clenched fists. They also shake rattles and rock back and forth. These are some of the most primitive expressions of psychic life. They are an overflow of psychophysical energy, which involves rhythm and frequency. We adults do this too: when we're bored and have too much energy, we might jog a knee up and down.

Now if you give a name to these rhythms: one . . . two . . . three . . . four . . . five . . . then you have numbers. So time, rhythm and number are connected at a deep level.

So you're saying that our sense of time evolves from our awareness of movement in space. This suggests a connection between time and space at an instinctual level.

We are all unconscious Einsteins. Children really only know that there is space/time!

How does this sense of time continue to develop?

When you analyse the dreams of Western children between the ages of four and seven years, you find (and I must use metaphor here!) that they're still in the pond that the stork brought them from. They lag back in a dreamy timeless world and resist coming into our time-ruled society. And these dreams are about the destructiveness of time.

For example, a young girl I knew was having difficulty when she first went to school. She couldn't adapt to

school time, timetables, sitting still and the whole progress of consciousness that school demands. At the same time she was having a repetitive nightmare.

She dreamt that a row of geese was marching by and everybody who saw them fell down dead. Then a train rolled by, and everybody who saw that also fell down dead.

Now, the geese and the train were a symbol of the mortal danger of passing time. That is, awareness of time creates awareness of mortality. It's as if the girl's unconscious was giving her a little shock by saying to her: 'Look, this is the world you're born into. Time passes and one dies. Pull your socks up, accept it and get going!'

Do you believe in precognitive dreams?

There's a way in which the unconscious knows the future in each of us and prepares us for death. When we eventually get older, and our body starts to give in, we have forebodings of the coming decay in both waking life and in dreams. We see the passing geese and trains again. I've recently started to collect people's last dreams to see what their unconscious had to say about death.

One example is from an old woman. At 8.00 a.m. on the day of her death she described this dream to her nurse. It was night and she saw a candle on the window sill inside the room, but it had burnt down and was starting to flicker. In agony she thought it would go out and she'd be plunged into darkness. Suddenly there was a blackout. But then she saw the candle again, and this time it was on the other side of the glass, outside, and the flame was big and burning quietly. Four hours later she died.

The candle is a well-known symbol of life and death. We speaking of snuffing life out like a candle. This dream seemed to be saying to the woman: 'You feel that your life is coming to an end and it will be followed by nothing but a great darkness. But beyond the threshold the life process somehow goes on.'

You see our instinctual side believes in the continuation of the life force after death. The unconscious gives us comforting dreams. I've come across these kinds of dreams in people who are atheists and don't believe in life after death. There was one man who was furiously certain that death was *the* end, that with the ending of the body everything was finished. But even he dreamt that things continued, and this made him angry. He didn't like these dreams.

137

It just shows that we have no control over the unconscious and it will insist on trying to tell people what is going on.

In another of these dreams a man who was dying dreamt that he saw a tree on the edge of a cliff with its roots exposed, about to tumble into the abyss. Suddenly the roots gave way, but instead of falling, the tree hovered in mid-air, then drifted out over the ocean. It's a similar dream: the tree doesn't fall but simply goes elsewhere.

Now, I don't see these dreams as hard evidence that there is immortality, but as Jung said, if you believe your unconscious you're healthier than if you don't. It makes you feel better.

How do you account for the kind of dreams that I've had, about driving fast into brick walls, then waking up in a cold sweat at the anticipation of violent death?

Most dreams about death don't mean literal death. When you dream of real death you dream about it in quite different terms. Real death dreams have a deep philosophical tinge and are very impersonal. They are

dreams of going to the other shore, of getting ready for travelling, often westwards, when the sun sets. And they promise an ongoing life process.

How would you describe the difference between time in dreams and time in waking life?

In our so-called civilised Western societies, time is linear and accurately measured. Even if we don't look at our watches, we still have a linear feeling for time and can often guess how many minutes have passed. This obsession is our primary experience of time in waking life. But in our unconscious, the place of dreams, time changes its rate of flow. It even stops.

How would you describe the unconscious?

It's a borderline term and you can't describe it *per se*. I can only do it with images: it is paradise and hell, the starry sky and the depths of the sea; it's a country where the railway tracks meet in eternity.

Yet it's a land where most of us live all the time.

We live with one foot in it all the time, and each night we plunge right into it. So we have this feeling of familiarity with it.

It seems that you're not too keen on Western linear time . . .

No, but realistically we have to live with it. Our whole civilisation is built on it. Many people have a conflict with it. Creative people, for example, often find it hard to be punctual. And I often feel like the young girl who had the nightmares. I don't like adapting to clock time.

Yet you come from the country of the watchmakers . . .

I'm a paradox! But unfortunately I've learnt to be punctual.

Do you think our obsession with time bodes ill for Western culture?

It's a symptom of our overestimation of the rational world. When I first went to Egypt we had a private driver who was supposed to arrive at 8.00 a.m. to take us out for the day. But he didn't turn up until 11.00 and the professor we had as a guide was furious. He asked him why he was late, but the driver just yawned and said he'd overslept. He saw no reason to apologise and at first we were indignant. But after a while I longed to live in such a civilisation. Life and people are more agreeable that way.

How is it possible to overestimate reason in such an unreasonable world?

In Western civilisation the rational is usually identified with what we call the solar or masculine principle, and this involves a striving for clarity and precision. But today this has reached the point where it's beginning to be negative. The medieval alchemists foresaw this when they spoke of the devilish black sun: they saw that the solar principle has a shadow, a negative aspect. We're only just beginning to realise this now ... that too much consciousness destroys life. That consciousness ought perhaps to be dimmed a bit to allow life to blossom. So today we find the feminist movement looking more to the lunar principle; that is, a woman's way of talking which involves less of the solar principle.

We're also finding this in science. In quantum physics it's no longer possible to be absolutely certain of anything below a certain size. Precision is good in certain areas, but we must understand the limitations of rational consciousness, recognising where it's useful and where it's not.

We Jungians also think that you shouldn't be too precise in psychology. It's another field where you can't apply mathematical values.

Is the sun masculine and the moon feminine in all cultures?

No. In Japan and Russia, for example, the sun is feminine. In ancient Egypt they associated the moon with the bull, and the bull with kingship, in spite of the menstrual cycle which you'd expect would make them think of the moon as feminine.

Psychologically, for the Egyptians, the ruling principle of consciousness was moonlight. That's to say they still lived in a dreamlike state of consciousness. It wasn't until about 3,000 B.C., with the invention of measurement and writing, that the lunar king became a solar king. And this was a big jump forward in the development of clarity of consciousness for mankind.

Don't you think that identifying women too closely with dreamy, intuitive, lunar values and men with rational, hard-edged, solar values boxes us in to our respective sexes too much?

An actual woman is not only feminine but also masculine. And an actual man is not only masculine but also feminine. We must always distinguish between what is feminine and masculine in the same individual.

Do you think men and women experience time differently?

In our culture men are expected to go out to work in factories, offices or wherever, places where everything is run according to clock time. Men's sense of time has become identified with this. Whereas, until recently, women have lived more according to their 'feeling' rhythm than clock time. Nowadays women are going out to work more and more, but those who stay at home can keep a little bit more freedom in the sense of being able to live more according to life's rhythms.

But life's rhythms at home often involve demanding housework and childcare, in isolation and unhappiness.

In general, relegating women to the home is regressive. But if the masculinisation of women goes too far it makes them neurotic. The refemininisation process can be healing. But if that goes too far a woman can become a dull sack of potatoes. It's all a question of temperament and combining the right elements. Some women are happier with a certain amount of masculine life, others are not. The feminine side of a woman might enjoy being at home while her masculine side might be very bored and longing to go out to work. These masculine needs have to be satisfied. It's always a question of finding the right compromise.

And the same for men?

The same for men.

Do you think that menstruation makes a difference to women's experience of time?

141

A big difference. At the time of menstruation a woman is generally in a more dreamy mood. She is closer to her unconscious. Jung once said that women should have the first three days of their period off at home to get away from the masculine rhythm imposed by modern managements.

How do we find the balance between our masculine and feminine sides?

By trial and error.

But our society wants to put us into one pigeon-hole or another.

That's the catastrophe. The individual should be free to discover the right rhythm in which to live. As soon as you put people in boxes it's the end of everything.

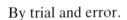

I was never asked properly, you see. There was no letter. Nothing definite. Nobody actually said, "Look! this is the situation. We want a drawing from you, right — by the last Friday in January or Monday. That gives you the weekend at least. Well, they did say, and I said "What?" and they said, I forget who it was who said, "Anything you like!" Well, that's a bit broad, isn't it? I mean, you try to do something — about TIME and make it profoundly whatever — off the top of your head. It was all a bit vague, so I put it off — well, I forgot actually, then somebody ● mentioned it again — TIME and I said, "How long have I got for it?" and they said 'WEEKS!!' so I thought "GREAT! I thought you wanted it tomorrow" so I didn't feel too anxious. Well, I forgot again, actually. Then, suddenly, out of ● the blue, Mike said "I believe you've done a drawing for the TIME book" and I said "Have I?" because I couldn't remember, and Mike said, "Well, the deadline was last Thursday." "Then I must have done" I replied, "Otherwise they'd have been down on me like a ton of breeze blocks. So that's O.K. I don't like letting people down — especially when I said I would — give them something, that is." Then the phone rang and its Ian and Irene screaming "Where's that TIME drawing you promised us?? It's late!!." And I said "I thought I must have done it because your deadline was last Thursday and I never heard." Ian bawled something about still having ten days. By now I'm confused and time's just whizzing by and I haven't an idea in my head and that was at least a week ago when he rang .... I'll do it tomorrow.

(Explanation ● under controlled conditions by Ralph STEADMAN about his TIME drawing 6 February 1885)

'Newtonian thought' – the mechanical logic of cause and effect – has underpinned Western science and technology for three centuries. Inspired by the philosophy of Descartes, enshrined at the heart of physics, it has seeped out into chemistry, biology, psychology and still further, by a process of theoretical osmosis. With the legitimacy that we accord science, this pattern of thinking, or paradigm, has come to dominate Western culture as a whole.

But this century has witnessed two revolutions in physics that have toppled this authority. Einstein's theories of relativity have disproved the absolute nature of Newtonian linear time. More significantly, quantum theory has proposed a universe in which the once cast-iron laws of cause and effect no longer hold complete sway.

Traditionally, the East has never invoked science as the arbiter of human experience. But in the West, the truth we grant to objective scientific knowledge has marginalised the intangible inner realities of being human because they are unverifiable. This section is about how key areas of science, largely through the two revolutions I have referred to, have come to relinquish the authority of certainty. In so doing, they have taken crucial yet vulnerable steps towards healing the immense breach between intellect and being, mind and body, that cripples Western sensibility.

Perhaps it's a measure of our dependence on the legitimacy of science that we should turn to revolutions in physics to authenticate the unprovable truths of human experience. Yet it is deeply reassuring when the citadels of causality and linear time are seen to be crumbling from within.

SCRIPT: K. WOLSTENHOME

1. SHOULD QUANTUM PHYSICS BE KEPT OUT OF SPORT?
2. COMPARE AND CONTRAST TRANMERE ROVERS WITH STENHOUSMUIR.

'In the heaven of Indra there is said to be a network of pearls so arranged that if you look at one you see all the others reflected in it. In the same way, each object in the world is not merely itself but involves every other object, and in fact is every other object.' Buddhist Sutra

If you drop two stones into different parts of the smooth surface of a pond, concentric circles of waves will spread out from each splash. Quite soon these waves will intersect, creating bigger waves where crest meets crest, and negating each other into patches of calm water where wave meets trough. The resulting criss-crossing network of wave and trough is known as an interference pattern. Light behaves in the same way.

145

TIME:
A Boffin Explains

IT IS MY THEORY THAT, BY AMPLIFYING OUR BRAIN WAVES TO THE POWER OF $v-t$ WE WOULD BE ABLE TO MOVE BACK AND FORTH IN PERSONAL HISTORY.

A laser beam is a very pure kind of light in which all the waves are of the same frequency. Imagine two laser beams converging on each other from different sources. At the place where they meet, an interference pattern of light and dark waves is created, analogous to the waves and troughs of the intersecting ripples on the pond. Now imagine that before meeting, one of the laser beams is diffracted (reflected) from the cross-section of a nautilus shell. When the beams now meet, an extremely complex interference pattern will be produced, which can be recorded on a photographic plate.

If a laser, or focused beam of light, is subsequently directed at the processed photographic plate, a three-dimensional image of the nautilus shell will appear to hover in space each side of the surface of the plate. This is a hologram.

The hologram of the spiralling nautilus shell on the front cover embodies the concepts of time which we are exploring in this book. The spiral is a visual symbol of time which is neither linear nor cyclical, but a dynamic synthesis of both that combines forward movement with the cyclical contingencies of life.

The spiral also has its history coiled within it, making present the traces of its past. In a sense, the whole development of the shell is present so that no one piece of it makes sense without all the others. Then and now, this bit and that bit, ask to be read as a whole in which space and time are interrelated aspects of the same unity.

The holistic properties of the spiral also characterise the hologram (the word declares as much). If a holographic plate is broken, the entire image can be reconstructed from any single fragment. The whole is enfolded in each part.

Although strictly speaking the hologram does not enfold development in time, none the less it has recently come to serve as a metaphor for the structure of the universe proposed by the physicist David Bohm, in which all parts and all time constitute an interrelated whole.

From most angles the hologram of the shell looks like a piece of silvered plastic. The image of the shell only emerges when it is hit by the right kind of light. Until then it exists only in potential as an invisible swirl of submerged light frequencies. Likewise, David Bohm has proposed an underlying reality in the universe, which he calls the Implicate Order. Like the hologram, this is a realm of frequencies and potentialities that underlies our illusion of concreteness.

Independently, the brain-scientist Karl Pribram has suggested that the hologram might serve as a model for the human brain. In his search for the site of memory he has proposed that memories may not be located in particular places in the brain, but spread everywhere throughout it in a complex domain of interacting frequencies.

Bringing these ideas together, Bohm and Pribram have speculated that human consciousness may be seen as the co-mingling of holograms with holograms in an ultimately seamless flux. The illusion of concreteness that consciousness gives us – our manifest reality of feelings, perceptions and life process – might be formed by the interference patterns created in the encounter between our holographic beings with each other and the holographic universe.

'Each person enfolds something of the spirit of the other in his consciousness.' David Bohm

Do you remember the little coloured stamps we used to get at Sunday school, each illustrating a biblical story, and how lovingly we'd stick them into the waiting spaces of the booklet? Do you remember the taste of the gum, and the guilt about last week's empty space that meant that you hadn't been there? Now, I can see in my mind's eye, framed by perforations ... David slaying Goliath, Absalom hanging from an oak tree, and Noah perched in his ark on top of Mount Ararat surrounded by the flood. There were so many tales of death and retribution.

There was a cast-iron drain grille set into the bricked yard at the back of the farmhouse. One very wet autumn it got blocked. This coincided with the failure of a washer on the outside tap, so for a whole day rain fell and water gushed unchecked into the yard, turning my play space into a lake.

This was a time, like many others, when the house was saturated with anxiety about the farm. Milk yields were down, the hens weren't laying well, and we had several tons of damp grain on our hands from yet another bad harvest. As my mother gazed helplessly at the bank statements, my father, I later discovered, tried to drink his problems away.

Noah and the Deluge must have been one of the Sunday school stamps at this time because I became convinced that the water outside the house would go on rising, and that we and all our animals would have nowhere to go. Long after the tap had been seen to and sludgy wodges of rotting leaves and chaff had been cleared from the drain, I walked through the yard in fear of drowning.

Perhaps these kinds of memories are so vivid, once recalled, because they are associated with difficult experiences that shaped personality. In a sense, the child is the interference pattern created by parental emotions, forming psychologically where the ripples of their love and anxiety intersect. Moulded and swamped, in this instance, by the field of their worries, perhaps I internalised their feelings, seeking form for them in the projection of my fear on to the image of the Deluge and the flooded back yard.

A bit later I became obsessed with that drain grille in quite a different way. Someone had given me a science book for boys from which I'd learnt about the shape and structure of atoms, and how electrons whirled round nuclei of protons and neutrons. The book also described what an atom bomb was, how it was more powerful than any other bomb, and could destroy whole cities, and went off when one bit of metal was fired against another like a kind of gun.

Not quite grasping the difference between plutonium and iron, I decided one morning to split an atom on that drain grille. Sneaking a heavyish hammer from the workshop I started bashing it on the drain cover as hard as I could, making a lot of noise, but failing to start a chain reaction.

I was very frightened at first, but after getting nowhere I overcame my anxiety about starting the holocaust and became distracted by the earsplitting bangs I was making. Watching the hammer as I swung it, I found myself wondering about the precise moment that it hit the drain grille. Slowing right down, I tried to imagine the point at which the hammer would be closest to the drain grille without touching it. I had a riddle on my hands, for surely the hammer could go on approaching the drain grille in an infinite process of deceleration. It could be for ever before they met!

In his book Wholeness and the Implicate Order, *David Bohm points the way to a synthesis of physics and experience, consciousness and science, in which we can begin to understand such apparently disparate themes as anxiety, particle physics and memory in relationship to each other.*

DAVID BOHM is Professor of Theoretical Physics at Birkbeck College, University of London. He holds a Ph.D. in Physics from Berkeley, where he has taught, and has also held positions at Princeton, the University of São Paulo and at Haifa. Dr Bohm is the author of *Causality and Chance in Modern Physics*, *The Special Theory of Relativity* and *Wholeness and the Implicate Order*.

Do you think astrology and the new physics are moving towards a common understanding of the universe?

It may be, but to prove the connection would require a tremendous research project, and I don't think anyone's willing to pay for it!

As far as I understand astrology, it's not the planets that cause what happens to us so much as forces which commonly affect both the planets and ourselves. The positions of the planets simply indicate these forces. To know exactly what these forces indicate – the meaning of the state of the planets – we must know our own past history. You see, through knowing the past, we see the meaning of the present better, and we get an idea of what to expect from the future, although not necessarily what *will* happen.

DAVID BOHM – MEMORY, ANXIETY

AND THE IMPLICATE ORDER

But at least I can safely predict that this time next year the earth will have circled the sun once.

I am not saying that there is nothing you can know about the future. The probability is high, assuming no object comes from outer space, that that's what will happen. Some day such an object will appear!

In the present you are remembering a number of past events in an order which leads you to expect a certain future. A lot of experience of that kind has been verified, so you count on it. But the future that we expect often fails to come. And our memories are extremely unreliable, except when accurately recorded.

But it must help to be concerned for the future? To want to predict it?

In one sense, it's valid, yes. For example, shall we prepare for changes between winter and summer climates? Shall we plant grain in expectation of future crops? These things are fine.

It's when it comes to more subtle events, like those of the mind, that it gets hard. Here, I think, we're in the same domain as quantum mechanics, where the attempt to predict will destroy.

Explain this a little.

I'm suggesting that there's an analogy between what's going on in the mind and what physics has discovered in quantum theory.

There is a principle in quantum theory called the Uncertainty Principle. In order to define something in quantum mechanics you must make an observation. But this observation will use energy. To look at an atom through an electron microscope involves focusing a beam of electrons on it. Now, the more precisely you want to observe that atom, the more energy you will need, until you reach a point where the atom you are observing will no longer be stable.

So the act of observation disrupts what you are observing?

Yes. In the case of an electron in an atom, the more precisely we observe its position, the less we can know about its momentum, and vice versa.

Now, you can compare this with human thought. If you try and think about a thought you are having, the thought goes somewhere else.

If you try to define the point at which a thought arrives, it changes its momentum. There's a kind of disturbance which we say also occurs when we try to observe an electron.

And perhaps it's the same with our emotions. The attempt to predict what we're going to feel like in the future is like the Uncertainty Principle. It will disrupt and change feelings. As in quantum mechanics, so it is in the mind: the more precise the observation, the greater the effect.

So you can't really separate the act of observing from the thing observed?

Well, as far as physics is concerned there is a grosser area where this separation is valid — for example, in everyday measurements like atmospheric pressure or the speed of a train. It's when you get down to a fine subtle level that the effect of the observation becomes significant, and with the mind it's similar. There are some very gross things which can be objectified, like predicting election results. And there are the fine subtleties, which are the most important, and these cannot be measured or objectified.

For example?   For example what makes you like, dislike, afraid or happy. The psychological issues that shape what people actually do. Here our concern for the future creates a serious psychological problem.

149

What is that problem?

Anxiety.

But surely anxiety has a basis in reality?

Yes, but it is largely self-created. We have created a society in which we give tremendous meaning to what we expect of the future and we get anxious if anything threatens that. But the future cannot be guaranteed, so if you give primary significance to what you expect from it, you'll inevitably get extremely worried.

Surely the nuclear threat is a valid anxiety?

Yes, but we ourselves have created it because we're worried about our futures. 'Will my country be safe? Prepare for nuclear war!'

Isn't it a natural human trait to worry about the future?

I think we worry about it far more than people who don't think so much about time.

We start worrying about the future when we learn about time. In the past human beings had a very loose idea of time. People thought of tomorrow as some vague period in the future.

Whereas in our culture, tomorrow is getting up, going to work, routine, schedule, timetable.

That's because of the way we organise our society. And again, this creates a serious psychological problem. It runs against the natural tendencies of the brain and nervous system to define time that sharply. It creates stresses in the system: anxiety and fear as to whether you're going to get things done on time, whether there will be time.

You see, I think time has become a kind of conditioning for all of us. We accept certain ideas of time tacitly, without question. We mimic what other people think about it.

What are these ideas that we accept without question?

The Newtonian concept of time which sees it as a uniform linear process: one moment succeeding another and each moment being the same moment for the whole universe.

And this is the 'common-sense' idea of time?

Yes. Assuming that 'now' is the same moment for you, me and the person in Australia. Yet our psychological experience of time is very relative. At a certain moment it's hard to tell what we mean by 'now'. If you're completely absorbed in something, 'now' may last a minute before you realise that any time has passed. And in a short dream a long time may seem to elapse, time sequences seem to overlap, the past seems present.

Yet we generally discount all this and say that physical (Newtonian) time is really absolute and universal.

Q: This reluctance to question these ideas has been seen as the limitation of the 'Newtonian Paradigm'. What is a paradigm?

DB: Thomas Kuhn brought the word in. What he meant by it was a set of ideas that scientists, especially physicists, have tacitly absorbed through their 'apprenticeship', without knowing they have absorbed them. And it implies they will feel very uncomfortable with anybody who changes these ideas, even fight to hold on to them.

Literally, a paradigm is a typical example. Physics, for example, which has a more defined paradigm than any other science, is taught in terms of typical examples. Learning physics is like an apprenticeship: you watch what the master does, picking up your ideas that way.

Q: Rupert Sheldrake calls this 'mechanical thinking'.

DB: Yes. Where you learn by mimicking others. You don't specify your assumptions, just tacitly pick them up. And these are the ones that are hardest to deal with.

$A_\nu + \frac{1}{3} A^\mu k_{;\mu}$

Vary wrt

$\delta \sqrt{-g} = -$

$\delta R_{\mu\nu} = \delta$

$\delta \Gamma^\lambda_{\mu\nu} =$

$\delta \Gamma^\lambda_{\mu\nu,\lambda} =$

$\delta \Gamma^\lambda_{\mu\lambda;\nu} =$

$\delta R_{\mu\nu}$

$^{\mu\nu} \delta R_{\mu\nu} = $

δR

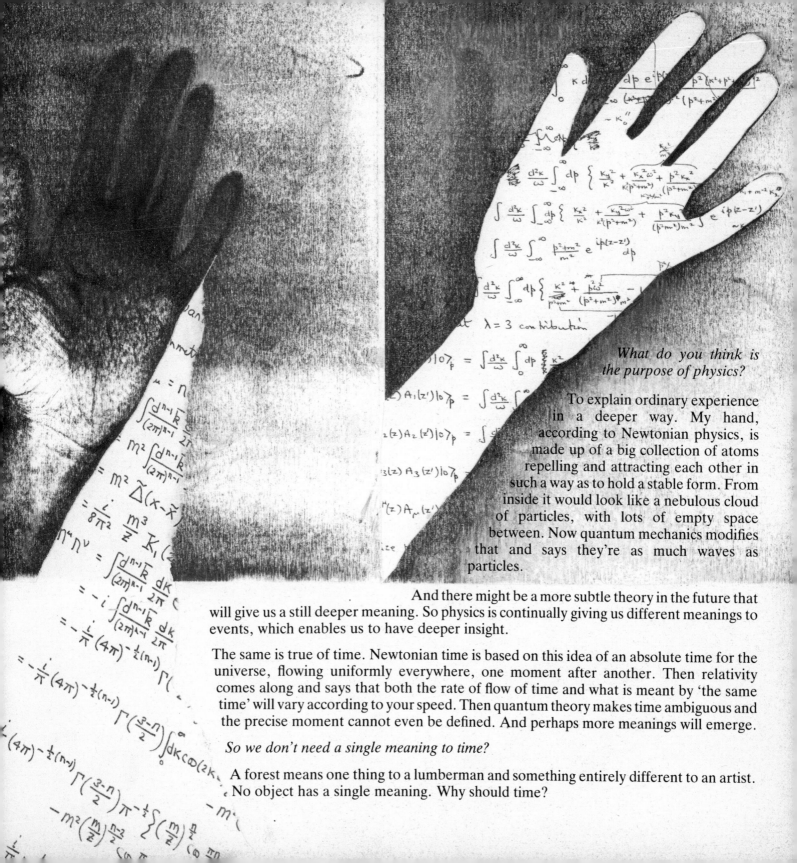

What do you think is the purpose of physics?

To explain ordinary experience in a deeper way. My hand, according to Newtonian physics, is made up of a big collection of atoms repelling and attracting each other in such a way as to hold a stable form. From inside it would look like a nebulous cloud of particles, with lots of empty space between. Now quantum mechanics modifies that and says they're as much waves as particles.

And there might be a more subtle theory in the future that will give us a still deeper meaning. So physics is continually giving us different meanings to events, which enables us to have deeper insight.

The same is true of time. Newtonian time is based on this idea of an absolute time for the universe, flowing uniformly everywhere, one moment after another. Then relativity comes along and says that both the rate of flow of time and what is meant by 'the same time' will vary according to your speed. Then quantum theory makes time ambiguous and the precise moment cannot even be defined. And perhaps more meanings will emerge.

So we don't need a single meaning to time?

A forest means one thing to a lumberman and something entirely different to an artist. No object has a single meaning. Why should time?

I start from a view of the universe as a whole that divides relatively into sub-wholes. And the basic sub-whole is the moment. But a moment may be of different kinds. It may be a symphony, which has movements, each of which are also moments. (The word moment comes from movement.) Or there is the 'moment in history' which might last for a hundred years. Whereas the moment of an atomic particle might be very very short. Moments are relative things and they are all in relationship with each other, some containing, some contained by, others.

The present moment, for example, contains, or enfolds, all past moments.

It reminds me of the layers of geological strata.

Yes, but even more mixed up. Each moment contains the enfoldment of the previous, and so on. And this holds true for all kinds of moments, the moments of consciousness as much as the moments of subatomic particles. A series of enfoldments within enfoldments. A set of Russian dolls illustrates the principle. You open one and find another. But this is too simple because enfoldment is more subtle: the past is spread all through the present.

A nice example is a hologram. If you break a holographic plate into fragments, each fragment can reconstitute the whole image. In less detail than on the whole plate, the complete image is enfolded in each fragment. But the enfoldment of time is something we can only take so far with analogies. We have not yet produced a machine that will have one moment enfold the previous one in the way we do with memory.

A nice example is a hologram. If you break a holographic plate into fragments, each fragment can reconstitute the whole image. In less detail than on the whole plate, the complete image is enfolded in each fragment. But the enfoldment of time is something we can only take so far with analogies. We have not yet produced a machine that will have one moment enfold the previous one in the way we do with memory.

A nice example is a hologram. If you break a holographic plate into fragments, each fragment can reconstitute the whole image. In less detail than on the whole plate, the complete image is enfolded in each fragment. But the enfoldment of time is something we can only take so far with analogies. We have not yet produced a machine that will have one moment enfold the previous one in the way we do with memory.

A nice example is a hologram. If you break a holographic plate into fragments, each fragment can reconstitute the whole image. In less detail than on the whole plate, the complete image is enfolded in each fragment. But the enfoldment of time is something we can only take so far with analogies. We have not yet produced a machine that will have one moment enfold the previous one in the way we do with memory.

A nice example is a hologram. If you break a holographic plate into fragments, each fragment can reconstitute the whole image. In less detail than on the whole plate, the complete image is enfolded in each fragment. But the enfoldment of time is something we can only take so far with analogies. We have not yet produced a machine that will have one moment enfold the previous one in the way we do with memory.

moment

...all in relationship with

How does memory enfold time?

Memory and time are almost synonymous as far as experience is concerned. Memory is the way we know that time has passed. Without some record of the past there would be no way of discussing what is meant by time.

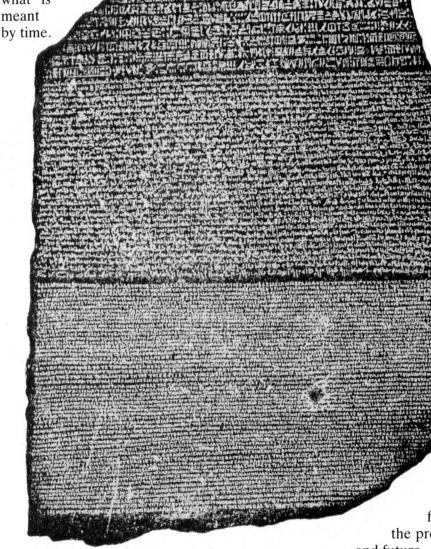

And as far as memory is concerned these records of the past are stored in our brains?

Perhaps. Also outside our brains, in libraries, computers, photographs. Or in the geological layers, or the rings of a tree.

They are always stored in this form that I call enfolded. The past is not stored directly, *as it was*. Newspaper print doesn't look anything like the events it's describing. So to learn about the past we must know how to unfold the storage and find its proper meaning.

Where is the past?

If we think of memory as the enfoldment of the past in the present, then all memory, whether it's personal, in the layers on the earth, or in libraries, is equally present now. We never have a direct experience of the past. It doesn't exist. It's gone, yet it's active in the present.

And the future?

. . . Never arrives. You might get very puzzled by this. The past doesn't seem to exist, the future is never here, and yet we say that the present is the dividing line between past and future.

ALLIES' DRASTIC ARMISTICE TERMS TO HUNS

The Daily Mirror

CERTIFIED CIRCULATION LARGER THAN THAT OF ANY OTHER DAILY PICTURE PAPER

TUESDAY, NOVEMBER 12, 1918

HOW LONDON HAILED THE END OF WAR

PLEASE PASS ON THIS COPY OR DISPLAY IT

The British Gazette

Published by His Majesty's Stationery Office.

LONDON, WEDNESDAY, MAY 5, 1926.

WORK AS USUAL.

TO-DAY'S CARTOON.

LEGAL ISSUE OF THE STRIKE.

Sir H. Slesser against Discussion.

THE TRADE DISPUTES ACT

AMONG THE MINERS.

Uneventful Days in Yorkshire.

MEN ANXIOUS TO RETURN.

No Trouble Expected.

ONE PENNY

THE BRITISH WORKER
OFFICIAL STRIKE NEWS BULLETIN

Published by The General Council of the Trades Union Congress

MONDAY EVENING, MAY 10, 1926

ALL'S WELL!

PRICE ONE PENNY

CHURCHILL'S STUNT

Daily Mirror
BRITAIN'S FIRST DAY OF WAR: CHURCHILL

JOHNNIE WALKER

Daily Express
Monday, June 1, 1942   One Penny   No. 13,107

The ruins of Cologne are hidden under a pall of smoke rising 15,000 feet after the first thousand-bomber raid in history

THE VENGEANCE BEGINS!

R.A.F. AT IT AGAIN

Cadbury means quality

ONE BOMBER EVERY 6 SECONDS, 3,000 IN 90 MINUTES

Sky over Cologne as Piccadilly

FIRES

DAILY EXPRESS
Opinion
Full and frank

THE CRIMEA DECLARATION
This is the full text of the Big Three communiqué issued in London, Washington and Moscow

DAILY EXPRESS
TUESDAY FEBRUARY 13 1945   No. 13,946

1 Germany is doomed. It is hopeless to resist   2 Occupation by Britain, Russia and France in four zones   3 Compensation to limit of capacity   4 Poland to lose

BIG 3: GERMANY TO PAY

VICTORY AND PEACE PLAN DRAWN UP IN CRIMEA:

Four Power control of Germany

THE THREE — First picture from the Palace in Crimea

Daily Express

No. 8127   LONDON, THURSDAY, MAY 13, 1926.

General Strike Call

T.U.C.

Plays records for 35 minutes automatically!

RADIO AND TELEVISION

popular table radiogram

Will America leave us?

The "special relationship" between Britain and the United States may be over, but is America really returning to isolationism? This month's meeting between Mr Wilson and Mr Nixon will be vital.

News Chronicle
LATE LONDON EDITION
WEDNESDAY, AUGUST 8, 1945

MAZO SAVES SOAP

"Hiroshima disappeared in cloud, boiling smoke and flame"

PILOT TELLS WHAT HAPPENED WHEN ATOMIC BOMB FELL

Last Allied warning: Yield, or we lay Japan waste

Yorkshire Miner
STRIKE ISSUE 9   NOV 1984

'Full confidence' in national leaders as coal-board unity cracks

THIS MAN IS IMPOSSIBLE !

VIOLENCE
You can't blame the miners for this

I HAVEN'T SEEN MacTHATCHER   YEP   NACODA BALLOT DON'T COUNT   NOPE   I WAS JUST GIVING A TASTER

THIS PLACE STINKS   WE'LL TALK   NUM SHOULD HOLD BALLOT   I DIDN'T SAY THAT!

WE'LL CLOSE TWENTY PITS   WE WON'T TALK

DAILY SKETCH, WEDNESDAY, JUNE 21, 1916.

OUR SUBMARINES IN THE BALTIC By RUDYARD KIPLING.   See P

DAILY SKETCH.

GUARANTEED DAILY NETT SALE MORE THAN 1,000,000 COPIES.   ONE HALFPENNY.

[Registered as a Newspaper.]

LONDON, WEDNESDAY, JUNE 21, 1916.

KING!"   The Same Old
In The Same Old

Second World War bomb damage to the British Library's Newspaper Library, Colindale

...if the present divides what doesn't exist from what doesn't exist, what shall we say about it?

Do you think your ideas have moral implications?

Although human beings are not very moral, we do none the less have morals which are based on the attempts of people to relate to each other properly. If people were entirely separate individuals, there would be no reason for morals. It would be everyone putting themselves first. The notion of morals implies the wholeness of humankind, and here the idea of the separate individual who isn't somehow grounded in the whole is just an illusion.

159

Quantum theory tells us that the universe is a whole and that the attempt to deal with fragments makes no sense. It's the same with people. It makes no sense to commit immoral acts in the long run. It's destructive all round.

But our culture encourages people to see themselves as separate individuals.

Yes, in the sense that we think only of our own future and not of the whole. Each person projects their own linear time into the future, towards the time when they think they will get what they want. And since everybody has a different idea of what they want, we all come into conflict. The consumer culture causes people to concentrate on things which won't satisfy them.

It has severe ecological consequences as well. It brings us into conflict with the planet.

We had tendencies towards fragmentation in the Stone Age, but they weren't dangerous then. Technological progress, however, has made this extremely dangerous. It means that humanity can't go on with a Stone Age mentality.

And a new mentality for you involves starting with a view of the universe as whole?

Yes. Everything enfolded in one whole, and unfolding from an infinite ground into finite moments. All moments enfolded within each other, and unfolding from a totality which is beyond time.

A source beyond time?

Out of which order emerges.

What is that source?

We can't say. But there are people who claim to have experienced the timeless.

So you're suggesting that the moments which constitute everyday time as we live it are projections from beyond time? This sounds very much like heaven or Eden. A place where God exists?

Well, that's what people may have called it in the past. But these words no longer mean anything and they tend to create a hostile reaction. We may fail to look at an idea because of our reactions to words.

I could rephrase the question in terms of physics: what happened before the Big Bang?

WHAT DOES THE QUESTION MEAN?

Time as we know it began with the Big Bang. There may be another kind of time beyond, in another kind of moment, within which this universe is being projected.

164

*Which would make everything we
know and experience just a moment.*

A moment in something much greater.

And that much greater something might be a moment . . .

You can't really follow it much further, can you?

*Do you think humanity faces some terrible learning process — the nuclear holocaust —
to reach what you're suggesting?*

I certainly hope not. A nuclear war and its possible
consequences might annihilate the human race. The amount of destruction would so
disorganise people that I don't think they'd learn anything.

What are you after in your work?
Ultimately, insight.
And physics offers this insight? Not only physics;
although physics has played its part. I don't really think you can separate
physics from the rest of life.

Some days when I'm very tired I take a rest late in the afternoon, and fall into a quality of sleep that's unlike any other. Instead of the semi-consciousness of cat-napping, or the steady trajectory of a 'good night', I sink very rapidly into a deeply relaxed dreamstate.

The grown-ups want me to cross the stream over that plank. They're urging me to crawl over that rotten sleeper to the other side, and I know I can't do it. 'Come on. Hurry up. It's easy. You won't fall. Oh, do buck up.' So I edge slowly, in terror, on all fours over the chasm of reedy water, rigid in the absolute knowledge that I'll drop and be swept by this little carrier into the spate waters of the main river a few yards downstream. And there in swirling brown vortices, weeds will strangle me, pike will bite me, eddies will suck me under. True to my fear, I tense and lose my balance, keeling over to the smiles of the tilting silhouettes on each bank. Falling, falling . . .

Opening my eyes, I see the clock on the bedside table but am unable to place it. I feel the familiar perspective of my bedroom, yet don't recognise it. I am suddenly awake from one of those deep two-hour sleeps; I don't know who I am, where I am, or what day it is. All the familiar scaffolding of my identity has deserted me. I'm in a no man's land, still burning with the imprint of that fast-receding dream, which I am forgetting because I can't find myself. I am dissolved in a timeless swell of nameless emotion. No, we. There is no I. We are simply process, striving to be me again. Groping, we search for familiar markers that will re-mind me who I am.

Now I remember. It's Thursday evening. It's dark because I went to bed at dusk. It's five to seven. I left the fire on. It's dinner-time. Am I going out? A snatch of the dream surfaces in a flash. Stay, stay. What am I doing on this river bank? What is this bad feeling?

Perhaps these timeless and formless states of being in the zone between waking and sleeping are the closest we come, short of meditation or the use of psychedelic drugs, to the preconscious states of infancy. That luscious stage where we were simply life process. A place where
166 *consciousness reluctantly emerged, crystallising around*

the grit of core experiences (like falling into streams): little islands of awareness forming in the flux as an indefinite sense of 'we' painfully coagulates into the certainty of 'I'.

The wonderful thing about quantum physics is how it confirms the intimate and complex experience of time revealed to us in dreams, memories and meditative states. In insisting that it's not ultimately possible to separate the observer from the observed, quantum physics tacitly gives us the confidence to value subjective human experience at least as highly as so-called objective knowledge. In so doing, it debunks the sole authority of Newtonian linear time – the time of timetables that we need to live by – exposing it as one concept among many: a single track marked by the Western mind across the boundless and synchronous expanse of all time that quantum physics uncovers.

When she had her first child, Danah Zohar found an extraordinary resonance between her understanding of time and process in quantum physics and her experience of her identity through motherhood. Before talking about time and motherhood, however, she outlines some of the basic principles of quantum physics which embody expanded concepts of time.

DANAH ZOHAR was born in the United States. She received her B.Sc. in Physics and Philosophy from the Massachusetts Institute of Technology in 1966, and completed three years' postgraduate work in Philosophy and Religion at Harvard University. She is now settled in London, and married to a Jungian psychiatrist with whom she embarked on motherhood at the age of thirty-eight. She is the author of *Through the Time Barrier*, a study of pre-cognition, and is a regular contributor to the *Sunday Times*.

DANAH ZOHAR – BEING, MOVEMENT AND RELATIONSHIP IN QUANTUM PHYSICS

1 – BEING

Quantum theory offers a very vivid description of the way that 'things', and by analogy 'beings', first come into and then subsequently pass out of existence. Such beginning and ceasing, or as I prefer, 'emergence and return', are summed up in the Principle of Complementarity, which states that matter (at the subatomic level at least) can be described equally well as consisting either of waves or of particles.

In classical, Newtonian physics, it was assumed that reality, at its most basic, unanalysable level, consisted of tiny, discrete particles which collide with, attract or repel each other. Wave motions, on the other hand (e.g. light waves), were thought to be the vibrations of some underlying jelly (the ether), not fundamental entities in themselves. So while the Newtonians used both the wave and particle descriptions of matter, particles were thought to be more fundamental than waves.

In this century, that assumption has changed. First, relativity theory abolished the notion of a mechanical ether, so that light waves came to be seen as things in themselves, basic units of reality. Then, more sweepingly still, quantum theory demonstrated that all matter behaves *sometimes* like waves, and *sometimes* like particles, depending on the overall conditions. The wave description and the particle description are complementary – like the right and left halves of the human brain, each kind of description supplies information which the other lacks. But an exact description in one set of terms precludes an exact description in the other set of terms. This is the nub of Heisenberg's Uncertainty Principle. According to Heisenberg, we are left with reality as a not fully analysable 'something', an indeterminate porridge of being that has both the particle and wave aspects of behaviour and to which we can never give any exact description independent of a given context.

At the level of everyday reality, we can see Heisenberg's Uncertainty Principle and the Principle of Complementarity as offering us a choice between different ways of describing the same system. For instance, we can think of waves as patterns of movement through the sea, or we can think of them as so many discrete, disturbed water molecules. We can think of a forest as a living system filled with all manner of things and crea-

tures, or we can break it down into trees, rabbits, flowers, etc. Still further, we can think of the chemical molecules in the trees, the individual body cells in the rabbits. Different kinds of things can be seen more clearly at the different scales, but who is to say which is more fundamental? Which, or what, more actually exists?

Going further still, quantum field theory points out to us that even those particles which do manifest themselves as individual beings are not permanent. At high enough energies, particles can be born out of a background of pure energy, exist for a transient while, and then dissolve again into other particles or into the background energy. Some of the individual particles' qualities are conserved (e.g. mass, charge, spin) but the number and type of particles is not constant. Such constancy is reserved for the overall balance of the entire system, that 'well of being' out of which the individual matter bits temporarily emerge and into which they eventually return.

2 – MOVEMENT

In classical physics, particles such as electrons were thought to move smoothly through space and time along continuous, unbroken paths. But quantum theory disrupted this ordered, predictable view of things by demonstrating that in

their particle aspect, electrons actually move from one position (energy state) to another in broken, jerky movements according to how many quanta (basic packets of energy) they have absorbed or given off. Worse still, according to the Uncertainty Principle, these jerky movements occur in an indescribable, unpredictable manner which makes it impossible for us to locate them in space and time as we know it.

Quantum physicists trying to describe the movements of electrons found themselves up against a fundamental difficulty: the harder they tried to scrutinise the speed and the position of the electron, the more elusive it became. The mere act of focusing on the particle was enough to disturb it. As the Uncertainty Principle sums it up: at a certain level of reality we come up against a barrier beyond which it is impossible ever to make a full set of exact measurements, and hence impossible ever to *know* exactly just how the constituents of matter are behaving.

The upshot of all this uncertainty is that quantum physics has had to resort to describing reality in terms of 'probability waves' – complex mathematical entities that sum up all the possible manifestations and properties (particle/wave duality, position, momentum, spin, etc.) of electrons which recede from focus whenever we try to look at them. Thus, probability waves give us an approximate picture of reality which is never more than a distribution of possibilities that, under any given set of circumstances, an electron will express itself in this way or that – and until it does so, *reality itself* (the reality of that electron) must be said to consist of probabilities.

Now the really mind-twisting fact about electron probability waves is that until a disturbed (energised) electron does indeed act out in actuality one of the manifold possibilities open to it as it gives off or absorbs energy within the atomic system, it behaves within the system as though it is acting out *all* of its possibilities, *all at once*. These possibilities, which behave as though they are smeared out across space and time, are known as 'virtual states', and they possess their own form of ghostly reality.

The situation for the electron is somewhat like that of a taxi driver who has just won the pools. His new-found wealth, he feels, makes it unsuitable to continue living in his modest council flat. A whole world of new possibilities has opened to him, and he wants to realise his greatest potential for ownership of his dream home. In the real world, of course, the taxi driver would have to explore each of his possibilities one by one, perhaps moving house a dozen times before feeling certain he had found just the right one.

But in the quantum world, the taxi driver would simply take up residence in *all* of his possible new houses *all at once*. If his bank manager wished to send him a statement, he would just have to send duplicates to each of the houses (since the taxi driver really is at all of these addresses). And if the driver wished to, he could stand on each of his various door steps and wave at himself.

In the end, having explored his various possibilities, the taxi driver would settle down permanently at one of his addresses, but not without having left traces of himself in the various neighbourhoods where he had temporarily taken up residence. The neighbours might remember seeing him and wonder where he had gone — some might even have changed their own life styles as a result of having lived next door to him. The taxi driver's 'virtual states' had possessed a kind of reality.

3 – RELATIONSHIP

Perhaps more than anything else, quantum physics has transformed our notions of relationship. Things and events once conceived of as separate, parted in both space and time, are seen by the quantum physicist as so integrally linked that their bond mocks the reality of both space and time. They behave as aspects of some greater whole whose 'individual' existence derives its definition and meaning from that whole. The once ghostly notion of 'action-at-a-distance', where one body can influence another despite no apparent exchange of force or energy is, for the quantum physicist, a fact of everyday reality.

It was Einstein (and two colleagues, Podolsky and Rosen) who first pointed out that the wave equations of quantum theory implied the possibility of action-at-a-distance. Believing in neither action-at-a-distance nor in the more metaphysical implications of quantum theory, Einstein thought he had shown up a contradiction which would in the end expose quantum theory as an incomplete description of reality. The alleged contradiction became known as the E.P.R. Paradox.

The gist of the E.P.R. Paradox can be understood by imagining the lives of a hypothetical set of twins, born in London but separated since birth, with one twin staying in London and the other going off to live in New York. During their years of separation, the twins have no contact with each other, indeed they are ignorant of each other's existence. Yet despite this ignorance and lack of communication, a psychologist studying the twins finds amazing similarity in their behaviour and life styles. Each has adopted the nickname 'Scotty', each chose to join the police force and rose to the rank of detective inspector, each prefers the colour blue, was married in the same year to a blonde named Joan, etc. How can we explain all this?

In what he called the Theory of Hidden Variables, Einstein suggested (following our analogy) that the twins must be identical twins, and that their shared genetic material had pre-programmed them to have similar lives. But quantum physicists argued that there was no such genetic explanation

— that all links between the twins simply followed from their being aspects of some larger whole. A physicist named Bell eventually settled the controversy by suggesting an experiment that was outlined in Bell's Theorem.

The basis of Bell's Theorem, staying with the analogy of the twins, was to suggest that the twin living in London be kicked down some steps, so as to fall and break his leg. No one could argue that shared genetic material would then lead the twin in New York to a similar fall, so he would either continue standing upright, thus disproving quantum theory, or he too would fall, disproving Einstein. In fact, when the twin in London is kicked and falls, the twin living in New York has an identical fall at exactly the same moment, though no one has kicked him.

At the subatomic level action-at-a-distance, or the correlation of events separated by thousands of miles as though those miles simply didn't exist — a hypothesis of quantum theory — is proven. Similar experiments have since been done to show similar effects across time — two events happening at different times correlate with each other in such a way that they appear to have reached across time in some synchronised dance. The extent to which this happens depends on to what degree a system is in a 'particle' or a 'wave' state. But the existence of such integrally linked relationships in quantum systems lends a kind of physical basis to Donne's conclusion that 'No man is an island, entire of itself'.

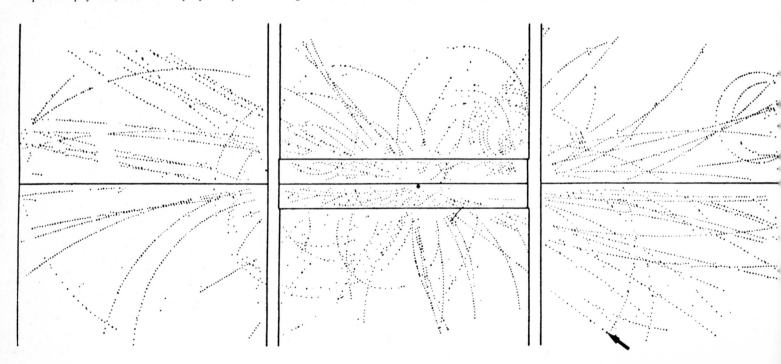

EVENT 2958. 1279.

Collision event from the U.A.I. experiment at C.E.R.

DANAH ZOHAR – QUANTUM PHYSICS AND MOTHERHOOD

How has being a mother affected your views on time?

Before motherhood I felt I was an individual person. More like a particle. I could focus on myself and say: 'I like this, I do that.' I had my projects and schedules and thought my individuality terribly important. But since motherhood I've found it very difficult to focus on myself. I'm much more a part of life's processes, giving this, taking that, simply being here. I've been involved with Anna's needs all the time. It's been much more like the wave half of the wave/particle duality.

I have also come to feel more closely related to the whole process of life around me: increased involvement with my parents and grandparents; with my baby's children and grandchildren. I feel myself part of this whole flow from way back to way forward, and this is the picture of reality portrayed in quantum physics.

The nub of quantum physics is really the Uncertainty Principle. The gist of it is that if you focus, say, on an electron in an atom, you can't find out both its position *and* its speed. The more you know about one, the less you know about the other. You can't abstract the electron out because it is part of the whole atomic system.

I feel very much more that way about myself now. Motherhood reflects the quantum picture of reality. My individuality doesn't have clear definition any more.

But I think it would be unfair to say that concentrating on our individuality isn't also an important part of being human, and I hope to regain that when the child's needs aren't so pressing.

What quantum theory has shown so beautifully is that MATTER IS BOTH WAVES AND PARTICLES. It's not fair to say that it should be either/or. I used to find it theoretically difficult to wrap my mind round this duality: how can things be both? Well I have a lived sense of that now.

What do we learn about time in quantum physics?

The quantum world is almost timeless. Things don't seem to happen in sequence. Everything is about relationship, ebbing and flowing together. It becomes unreal to talk about before and after: you can't say that this happened before that happened.

A good example is what are called 'virtual states'. Imagine the classical model of an atom where you have electrons whirling round a nucleus: now, let's say that a cosmic ray comes from outside and interferes with it. The electrons will be disturbed from their orbits and suddenly go crazy, jumping all over the place from orbit to orbit, shell to shell, high to low energy states, and so on. And there's no rhyme or reason to which way they're going to jump. Now, if we were to try and find the position of one of these electrons, we couldn't. They are all everywhere at once.

You mean that the same electron can be in different places at the same time?

A disturbed electron will seek its most stable position in an atom. And it doesn't do this by trying out this position, then that position. It tries them all out simultaneously. In this context, the electron is a wave and it spreads itself throughout the space of the atom. Being everywhere at once are its 'virtual states'.

But we live in sequential time. Up in the morning, off to work, home to bed. This makes time in the quantum world seem a long way from our experience.

Not necessarily. I think it's the kind of time that we live in our quieter moments. The time involved in dreaming, reflecting, even in being a mother. These experiences of time are not easily given to structure. When I was pregnant, schedules didn't sit well with me at all. Trying to structure the flow of experience into Newtonian sequential time gave me a headache.

The Sunne Rising.

BUsie old foole, unruly Sunne,
   Why dost thou thus,
Through windowes, and through curtaines call on us
Must to thy motions lovers seasons run?
      Sawcy pedantique wretch, goe chide
      Late schoole boyes, and sowre prentices,
   Goe tell Court-huntsmen, that the King will ride
   Call countrey ants to harvest offices;
Love, all alike, no season knowes, nor clyme,
Nor houres, dayes, moneths, which are the rags of time.

      Thy beames, so reverend, and strong
      Why shouldst thou thinke?
I could eclipse and cloud them with a winke,
But that I would not lose her sight so long:
      If her eyes have not blinded thine,
      Looke, and to morrow late, tell mee,
   Whether both the' India's of spice and Myne
   Be where thou leftst them, or lie here with mee.
Aske for those Kings whom thou saw'st yesterday,
And thou shalt heare, All here in one bed lay.

      She'is all States, and all Princes, I,
      Nothing else is.
Princes doe but play us; compar'd to this,
All honor's mimique; All wealth alchimie.
      Thou sunne art halfe as happy'as wee,
      In that the world's contracted thus;
   Thine age askes ease, and since thy duties bee
   To warme the world, that's done in warming us.
Shine here to us, and thou art every where;
This bed thy center is, these walls, thy spheare.

John D

Outings. Pram walks morning and
Habit training. Napkins should I
The child should be held out quiet
should wear napkins at night until c
Toys. As before and balls (sma
drum and sticks, pyramids, teddy be
lorry and bricks, horse and cart, p
leaves), wooden animals, pegs and h
Food—see page 113.

12 P.M.
Gravy, 1 oz., *or*
Broth, 2 oz.
or
Fish pudding, 2 oz.
Sieved potatoes, ½ oz.
Vegetable marrow,
Sieved spinach *or*
cabbage, ¼ oz.
Bread, ¼ oz.
Milk, 5–6 oz.
Sugar, ½ oz.
Water to drink

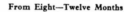

From Eight—Twelve Months

Feeding. 6 and 10 a.m., 2, 6 and 10 p.m.
Breast Fed. As before till 9 months, and solid food (see below).
Weaning usually commences at 8 or 9 months, for method see
p. 79.
Artificial Feeding. For milk mixtures see pp. 80 and 81.
Solid Food. Child should learn to hold spoon and hand be
guided to mouth. For details see page 112.

Routine
Bathing. 8.30 a.m., top and tail.
5.30 p.m., bath and cool sponge (see pp. 44 and 52).

ering and Play. 9–10 a.m., 1–2 p.m., 4.30–5.30 p.m. In
s on floor out of play-pen part of the time, to enable the
explore. Put in play-pen with toys for safety when left
ed.

Approximately 17–15 hours. All babies from 6 months
have 1–2 hours sleep in the afternoon as well as morning,
or cot.

gs. Can be propped up for longer times when out, but
be laid flat when tired or sleepy.

As before and hollow coloured blocks, large coloured
arge coloured rubber balls, drums (without sticks), small-
carts and toys on whe
themselves up, and ou
nselves.

From Twelve—

ake early sit on chambe
juice if wanted, or who
arrot if digested.
g. 8 a.m., 12.30 and 5 p
Food. Child should be
n mug. See page 113.

g. 7.30 a.m., cold bat
or method, see p. 46.

, warm bath—encourage to play with floating toys, etc.,
once weekly.

hours. At night uninterrupted for
lifting). By day rest and sleep may
l 2–3 or 3–4, or if not ready for
12.30–2.30 or 3 p.m. (dinner at

portant, even if the child does not
to be cuddled lying down if very

eal times, now all activity is play.
tre. It is advisable at this age to
so to crawl upstairs as long as they
m on hands and knees backwards,
s independence.

	TEA	SUPPER
	Milk, 6–8 oz.	Broth
	Sugar, ¼ oz.	3–4 oz.
	Bread	*or*
	Toast	Milk, 6 oz.
	Rusk	
	Plain	
	home-	
	made ⎱ Singly *or*	
	⎰ combined,	
	Madeira ⎰ 1–1½ oz.	

Outings. Pram or go-cart morning and afternoon.
Habit Training. Teach to wash own hands and fill own mug
from tap for drinking; to help fetch and carry, to put away toys,
to help dress himself.
Toys. As before, and should include large-wheeled toys, some
of which can be taken out of doors, *e.g.*, cart or lorry.

From Eighteen Months—Two Years

The general routine is as from 15 to 18 months. At this age
children enjoy being helpful and independent by fetching own
clothes and utensils, laying and clearing table, helping wash up,

Do you think that children come into the world with any sense of time?

Babies come into the world with their own rhythms, but not with a sense of the structure of time as we know it. We impose that structure and if we do it too harshly, the baby rebels. Initially they don't know the difference between day and night: they wake and sleep, sleep and wake. They have no notions of times of day. They just eat when they're hungry and sleep when they're tired. But gradually, as the child can take it, you begin to impose the first time structures. An eating routine. Then you try to make a distinction between day and night, with a little more noise in the daytime. Slowly you bring them into our world of order, which I don't think is natural to them but, if you do it gently enough, hopefully you don't rip them out of the underlying swell of their own rhythms too suddenly. If you do, they cry, because it's literally painful. I think it is to us too, but we just aren't conscious of it.

What are the consequences of ignoring these inner processes?

I think we cripple ourselves. If we're not in touch with them, we're not fully human. We lose touch with each other and with nature and life's processes.

But we can't do without structured sequential time.

No. Clearly we need it. We're kicking this Newtonia sense of time around as if it's a fallen being that's don something naughty. But it's foolish to say we should a drop out and just experience being. The baby woul collapse if she didn't get her breakfast roughly at nin and her lunch roughly at noon. But I think that in ou busy organised urban lives, structured time does tend t crowd out this other experience. To have a balanced li we need respect for both.

Do you think Jung's ideas about synchronicity have much in common with the picture of time that emerges from quantum physics?

Yes, very much so. Although synchronicity isn't a scientifically founded theory. It is a very good poetic description of action at a distance or action across a temporal distance, both of which Jung was concerned with, because of his interest in the I Ching.

The gist of synchronicity is that things that have a similar 'meaning content' will be drawn to each other, across space or across time. Jung was very interested in meaningful coincidences: if I think of someone I haven't seen for ten years while I'm walking down Oxford Street, and, suddenly, that person is in front of me, for example. Or with the I Ching, if I throw my three coins, with a sincere question in my head, uncannily, it gives me a relevant answer. It's as if I was actually consulting the Chinese sages who wrote the book thousands of years ago. Time seems not to exist.

This seems to raise important questions about free-will. It suggests that our actions are somehow predetermined in a universe where time (past, present and future) is all laid out.

No. It's more that all potential is laid out. To return to the disturbed electron in the atom seeking a new position. In the end, it will probably plump for the easy option, its lowest energy state. It will take the easy way out, as we usually do. But that doesn't mean that either we or the electron don't have the potential to take more difficult ways out. It's as though all possibility is there, and free-will is whether or not we wish to make use of these possibilities. The fact that most of us don't is neither here nor there.

We're coming to terms with these ideas through physics, yet in Eastern cultures they have been central to their philosophy and religion for millenniums.

Quantum physics made us more sensitive to the East and I think that to a certain extent Eastern mysticism and wisdom are creeping into Western culture via the physics door. It's pushing it a bit to say that everything in Eastern philosophy is the same as in quantum physics, but the general sense of time, process and relationship is similar.

173

For me,
part of the
poetry of quan-
tum physics is in
the sudden emer-
gence of particles
out of nothing. In an
atom smasher, for
example, particles sud-
denly erupt into being and
then just as suddenly go out of
being. At Anna's birth, I had
this overwhelming experience
of a new being. And it's the same
with death, which I haven't had as
much experience of: if you're at the
bedside of a dying person, suddenly you
feel that their being is 'going back' as it
were.

*You speak of 'going back' as if there were
somewhere to go back to?*

This is one of the things that motherhood
has changed for me. Before Anna was born,
  I was very afraid of death. I thought of it as
    a void, the end of everything. But
        Anna's birth was like the
           emergence of a being
              I clearly felt had
                come   from
                    some-
                      where.

174

Like a particle?

It's as if in dying you return
to the pool of being from
which you emerged in the first place.
And this is like quantum theory, where suddenly there's a
particle, and almost as sud-
denly, it's gone.

*To some order beyond
this world?*

Yes. To a well
of energy, out of which things
emerge into a particle reality, becoming
matter for a while before flowing
back into energy. David Bohm's Implicate Order is surely a
poetic expression of quantum physics. In the Implicate Order, all time is there,
enfolded within the totality. And as I was saying about free-will and the
atom, for Bohm the Implicate Order isn't all laid out. It's a vast
sea of potential, a dynamic process.

An ocean?

A vast well of being and potential, the totality of which we all spring
from, emerging, flowering for a time, and then going back.

This swirling oceanic place sounds very mystical.
I think the Implicate Order is a very good description
of what all the world's mysticisms
have talked about. I don't know whether Bohm meant
this to be a religious theory, but
it certainly is one I think

Do you agree with it?

Yes. And I think the important thing is what we do during our temporary sojourn in world time. I think we have a moral responsibility to return whatever it is we give back to the well of being in a slightly richer state than it came to us in. This is what morality is about. We emerge to make something out of life while we have the chance to live it in an individual state, so that whatever it is we return makes the fabric of being just that bit richer.

So in a way we are helping the well of being (some would call it God) to evolve?

I think of us as God's partners in evolution. He needs us on the world stage. We think we are carrying out our seemingly petty temporal acts, but in fact we are weaving this enormous web.

And the electron in the atom, the baby in your life . . . both of these are manifestations of events that emerge, act and return?

Yes.

Atomic physics has given you these insights. It's also given us the bomb.

I don't see that as a contradiction. Life is a sea of potential. Both the bomb and the insights we're talking about are potentials relating to atomic physics. It's like the genie who offers you three wishes: they can be for good or for evil. He just offers you wishes. We were offered the atom and all its insights. If we choose to have a bomb out of this, it is just some of the evil we've brought upon ourselves, and we're not doing much for the web of being in my opinion.

So we're living the dark side of our free-will?

We very often do, don't we!

The journey inwards towards our point of origin is the perilous business of psychoanalysis. The therapy of it lies in handling what is revealed and re-lived on the way.

This process of regression, in which one veil is lifted from another, and another, seems theoretically endless. No sooner have we been enlightened by recalling, say, a repressed trauma of puberty, than we discover it is formed, like an onion skin, round a trauma of childhood, which in turn is given shape by a similar pattern of events in infancy. So where do we stop? At the age of three, two, one? At those critical stages (if we know when they are) when ego is being formed? At the immense and knotted difficulty of our own births? In the ebbs and flows of gestation? At the very point of conception?

Different kinds of psychotherapy have their different positions on all this. But if our quest is to get a glimpse of 'the well of being from which we all sprang and which we will all return to', perhaps for our purposes, the further back and in we can see the better.

But isn't there a natural barrier that prevents us reaching the events of our preconscious history? It seems like a contradiction in terms to gain consciousness of a time before our egos were formed.

One way of dissolving this contradiction is to let go of the conviction that our identity exists as something individually distinct from the identity of others. Then we might travel deeper, through the threshold of the formation of our own ego, encountering the spirits of our parents, siblings and friends as they moulded the raw stuff of our own natures.

And we'd encounter the terrifyingly composite nature of our personalities, stitched together where the webs and fibres of other people's personalities intersect. Then, unravelling the collective patchwork, we'd go still further back, dissolving into the collective effort of our own evolving organism. A massively energetic collaboration of rampant, splitting, cellular growth coming wholly together. And plunging still further back down the corridors of mitosis perhaps we'd stumble on the first cellular division, even the spark of fertilisation itself. Where will it end? When will the mystery show itself?

178 These kinds of journeys, like Dante's in The Divine

Comedy, feel hazardous because they involve the voluntary disintegration of personality. The terror lies in the question: will I get back in one piece? And who can say?

Never mind. Follow me now on a day-return, no-risk fantasy. We'll be like test pilots breaking the sound barrier for the first time, but today we'll fly faster than the speed of light and catch the elusive secret. Let's go! Like Chuck Yeager in his X-1 rocket plane, flashing back through the events of childhood, ever accelerating inwards and backwards, through the womb, to molecular stuff, to the particle dance. Now the turbulence starts. A little bumpy at first, then buffeting this way and that as we go faster and faster. Hold on tight to the stick, and don't count on the going getting suddenly smooth as it did for Chuck Yeager when he went through the sonic wall without so much as a bump. We're flying straight into white light, 'pushing the envelope', as the shaking and shattering starts. How heavy we are, and how much effort it needs to push us that little bit faster, into the light that dances elusively ahead of us, keeping its distance. If we could just, if we could, if we could just touch . . . I can't hold it . . . I have to let go . . . Down into a glorious battering that splinters me into a thousand silvery entities. All those egos, like so many fragments of a hologram, mustering resources to hurl ourselves at the mystery just once more, only to be fissioned and multiplied yet again.

Now we are simply a soft ocean. There's no you and no me. Just us, undulating in pinkness, bathed in white light, with no sense of beginning or end. And we can see the faint outlines of a human face taking form behind the brilliance of that last inpenetrable veil. We can see the smile.

This kind of inwards psychic journey is complemented by the outwards journey, through the solar system out into the galaxy, and beyond, to the seemingly boundless tracts of the universe. There, on the brink of black holes, or in the unimaginable energies of quasars, we are faced with the same conundrums.

PAUL DAVIES is Professor of Theoretical Physics at the University of Newcastle upon Tyne. He is a frequent broadcaster and author of many books, including *God and the New Physics*, as well as *Space and Time in the Modern Universe*.

PAUL DAVIES – NEWTONIAN TIME, EINSTEIN, EVENT HORIZONS AND BLACK HOLES

The Western 'common-sense' idea of time, linear time, is often referred to as Newtonian time.

I think Newton, along with Galileo, was the first person to think systematically of time, like a mathematician. Before Newton, most people's image of time was in the sense of organic rhythms, the seasons, cycles of birth and death. Certainly as something which could be used to gauge or catalogue events, but not something precise in the mathematical sense. Well, for Newton it was crucial to use time as what we call a parameter in his theories: something which has precise values and which is used as part of a system of equations.

Now, before Newton people saw distance (space) in a mathematical way: the Greeks had a beautiful geometry of angles, distances and measurement. But the idea of time entering into the proof of an equation didn't occur to anybody before Newton. He put time into his equations and made it a central feature of his dynamics. He was using time to describe motion in a very precise way.

So for Newton, time becomes something that has an almost objective quality to it. It's as if it's more 'out there'.

That's the whole point. The idea in Newtonian science is to become more objective. Of course time is a very personal experience, and the need was to abstract away from that and find a way of thinking about time in the laboratory, or 'out there', as you say, in the universe. In putting time on to this abstract mathematical footing, Newton spoke of 'absolute, true and mathematical time'.

What does that mean?

I think Newton was the first person to state that there was a universal public time which had this precise mathematical quality of 'flowing equally'. That was the terminology he used.

179

We now know that this is wrong. With the theory of relativity we know that time is more personal in the sense that it's related to one's state of motion.

How does Newtonian time go wrong?

Well, it works very well under a wide range of circumstances, like catching trains, watching T.V. programmes, even flying spacecraft to the moon. But it begins to go a bit wrong when we want to do very sophisticated things like accurate satellite communication and navigation.

When a system, be it a clock, a spacecraft or a subatomic particle, is travelling very close to the speed of light, Newtonian ideas of time go badly wrong.

And this is the central feature of the revolution in physics that occurred at the turn of the century. And, unusually, it was very much the work of one man – Albert Einstein.

The revolution came about because of a contradiction that had been developing in physics. On the one hand there were Newton's ideas of space, time and motion, which had been around for 200 years and had worked very successfully. On the other hand there was the more recent work of James Clerk Maxwell on the motion of electromagnetic waves.

At the same rate everywhere?

Yes. The idea was that although our subjective experience of time could vary, there exists something called an ideal clock. You can imagine lots of synchronised ideal clocks in various regions of the universe, all keeping some sort of precise absolute time.

In Newton's mechanics, speed is relative. If we say a car is moving at 30 m.p.h. this speed must be relative to something: to the earth, which is moving round the sun at such and such a speed, which is moving round the galaxy at such and such a speed and so on.

Now, Maxwell had a system of equations which described how electromagnetic waves spread out like ripples from a stone thrown into a pond. But these ripples spread out through apparently empty space: the warmth and light of the sun – its electromagnetic radiation – is propagated over 93 million miles of nothing. And Maxwell's equations also gave a definite speed for these waves: about 186,000 miles a second, which is very fast.

For the physicists of the nineteenth century these notions were problematic and paradoxical: if these waves moved through empty space, what was their speed relative to? They'd always thought of waves as undulations in some sort of medium, like sound waves vibrating and carrying in the air. So now they assumed that there must be a medium for the sun's electro-magnetic waves which their speed was relative to – a tenuous invisible something that filled what we think of as the empty space of the universe.

This was all right until they tried to measure the speed of the earth through the ether. Now, although this speed never appears anywhere in Newton's mechanics, they believed that just as fish swim through water at a certain speed, so the earth must move through the ether at a certain speed. So they devised experiments, using light signals, to measure this speed. The most famous of these, conducted around 1890, was the Michelson and Morley experiment which found nothing. The earth was apparently at rest in the ether, which was clearly non-sense because it was at least going round the sun. There was obviously a contradiction!

Then a few years later, along came Einstein.

Einstein couldn't accept that the earth was somehow miraculously static, and proposed scrapping the ether idea altogether. Instead he suggested that there was something very wrong with our notion of space and time and motion, particularly in relation to light. And he made this suggestion, which at first sounds nonsensical: that the speed of light that Maxwell found – 186,000 miles per second – is always the same for everybody, irrespective of how fast they are moving.

Now this just seems absurd. If I make a sound, and run after it, it will be receding from me at the speed of sound minus the speed at which I'm running. If I run after my sound at the speed of sound, I'll catch it up. If I run after it faster than the speed of sound, I'll go through the sound barrier.

But Einstein was saying that when it comes to light, all this is badly wrong. If you have a little pulse of light receding from you and you rush after it, you will never gain on it, however fast you go. And the corollary of this is you can never run faster than light.

181

Since Newton, everybody had thought that you couldn't monkey around with space and time. They were fixed. Whereas the speed of light would change according to how you were moving.

Einstein turned all this round. He said that the speed of light was something fixed in all reference frames for all observers irrespective of how they moved. And he reconciled this with 'common sense' by saying that it's space and time that get distorted and changed according to your motion.

Let me clarify this for myself. Suppose I'm moving at 90 per cent of the speed of light and I am pointing a torch so that the beam is in the direction in which I'm moving. Now is the speed of that torchbeam my speed (90 per cent of the speed of light) plus the speed of light?

I know a better way of handling this. Suppose we had two very efficient space rockets which could travel at 90 per cent of the speed of light and we sent them off from the earth in opposite directions. Now surely the observer on one rocket would see the other rocket receding at 1.8 times the speed of light. Faster than the speed of light. Well, the answer is no. It's true that on earth we would see each rocket recede at 90 per cent of the speed of light, but each observer on each rocket would see the other rocket moving at something like 99 per cent of the speed of light.

In arithmetic $1 + 1 = 2$. But in relativity this is not true. The way you add velocities together is such that, however many you add, you will never quite reach the speed of light.

Now we have particle accelerators that boost the speed of subatomic particles. They go round in a ring, and each time round they get a kick. But however many times and however hard you kick them, you never quite reach the speed of light.

Where does all this energy go?

It's converted into mass. It's as if the particles are resisting approaching the speed of light by getting very heavy and ponderous. At these speeds, to make them go even one mile an hour faster requires a tremendous amount of energy.

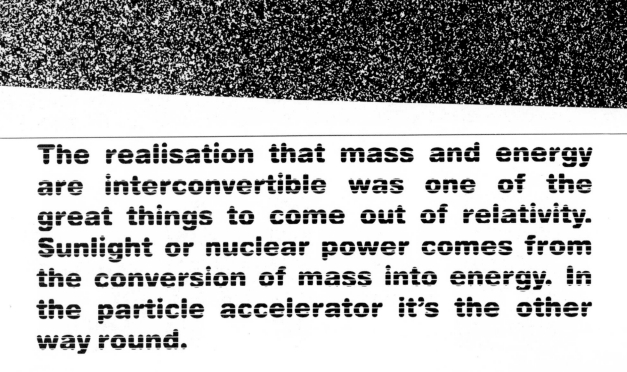

The realisation that mass and energy are interconvertible was one of the great things to come out of relativity. Sunlight or nuclear power comes from the conversion of mass into energy. In the particle accelerator it's the other way round.

$$E = Mc^2$$

And what happens to the time of our particle in the accelerator?

At close to the speed of light, their time-scale is stretched, or dilated, up to twenty or thirty times.

So subatomic particles live longer at these speeds?

They live longer in our frame of reference. If you could travel with them, their life time would appear normal.

If I could see a person flashing past me at the speed of light, that person would appear to be living in slow motion.

Yes. Likewise with a clock: its hands would be moving more slowly than your clock.

And how would a person flashing past me at the speed of light see me?

Exactly the same. It's a reciprocal effect. Each of us would see the other living in slow motion.

Suppose you and I were twins and I got into a very fast rocket and went off to some nearby star, and came back ten (earth) years later. Now, if I travel close to the speed of light, what would have been ten earth years for you may only have been five earth years for me. So when I get back, I have aged less. I am five years younger than you, even though we are twins.

Now the reason that I am the one that has aged less is that I have been doing all the manoeuvring: I've accelerated, gone off, turned round while you have just sat there.

Our time-scales have got out of step. It's not that I can increase my lifespan in my own frame of reference by dashing around like this. It is only relative to you that I will live longer.

So time travel is possible?

Yes. If you were the twin in the rocket, you could reach the year 2000 in say, two years, whereas for me, left behind on earth, it would take fifteen. Time travel in that sense is possible, if you had a powerful enough rocket.

And I'd only be two years older to your fifteen?

That's right. You can visit the future. But you can't come back again. You must stay there. You can't have somebody drop dead before you fire the gun.

This implies that time flows in only one direction?

There is a fundamental asymmetry to events in time. A lopsidedness. We see this irreversibility all around us. People get older and die. Things change in such a way that they can't change back. We express the fact that things can't get younger by a quantity called entropy. Entropy is like a measure of disorder, or chaos. The universe as a whole is slowly dying or crumbling. Things left to themselves tend to run down. The entropy of the universe, its disorder, is rising.

This is the Second Law of Thermodynamics.

Now, you may think that there are examples where new order and structure arise spontaneously. Surely a newborn baby is an increase in order? But if we look a little closer, we see that we are paying for this increase through the prodigious burning up of the sun's fuel. There are little islands of increased order – life – where the entropy is going down, but when you draw up a wider balance sheet, it's actually going up.

Now, this implies that the universe started with a stock of order that is progressively running down, and the question is immediately raised: how did the order get into the universe in the first place?

Well, some people used to think that if the universe is getting more and more disorderly, then God must have created it in a condition of order, and this order is simply running out. Today we think of it rather differently.

Looking back, it seems that the Big Bang, the origin of the universe, was rather a disordered thing. Lots of stuff spread around with no structure, no atoms, no galaxies or stars. None of the present organisation. Just primeval chaos and disorder. Now if the Second Law of Thermo-dynamics says that every day the universe gets more disorderly, how did this original disorder turn into the order that subsequently became disorderly? We're faced with a paradox.

The answer lies in the expansion of the universe. It is not just a fixed thing which sits there and runs down like a clock. It is more as if, in the first few minutes of the Big Bang, the universe was expanding very rapidly and this acted rather like a winding mechanism, putting the order in.

But doesn't the Second Law also apply to this energy needed to expand the universe?

No. The Second Law doesn't apply to an expanding universe.

Can we travel backwards in time?

No. You can slow up time by an infinite amount. You can even bring it more or less to a standstill. But you can't reverse it.

How can time be halted?

This is what happens with black holes. With relativity we've seen that time can be stretched or dilated with very rapid motion. But gravity will do it too. For example, it's been verified that a clock on the surface of the earth runs fractionally slower than a clock out in space. The difference is small because the relatively small size of the earth means its gravity is weak. If we were to place our clock on the surface of the sun, the difference would be much greater. Now, if we took the sun and shrank it to the size of Newcastle, we'd have an object whose surface gravity is so enormous that a teaspoonful of matter weighed there would be heavier than all the continents of the earth. This is a neutron star. Such things exist and we see them in the sky. Due to their enormous gravity, time on their surfaces would be ticking over maybe two or three times slower than here on earth.

If you imagine shrinking the neutron star still further, the time distortion gets increasingly out of step with earth time. There comes a point, when its radius is about a quarter of that of a normal neutron star, when this time discrepancy begins to escalate, shooting off to infinity. You reach a kind of infinite time warp where a clock on the surface of this shrinking neutron star would appear to us on earth to be completely frozen in time. If we could see it, that is, which we couldn't, because one of the other effects is that the frequency of light waves is slowed down. So light leaving the surface of such an object is stretched and slowed to a point where it becomes very dim, and dark.

188 As the neutron star continues to collapse under the force of its rapidly increasing gravity, it shrinks down to nothing, leaving a hole, a region around it which is empty because the star has collapsed out of existence. A black region, because dim light has been reduced to blackness. A black hole, like a prison, because nothing can escape – not even light. It is a region of space which, relative to us, is beyond the end of time. For this reason, if you were in a black hole – as with light – you could never get out because you would have already travelled beyond eternity.

So black holes are more than hypothesis?

Yes. Indeed, the stretching of time by ordinary stars has been known for many decades now. You can measure it if you look at the quality of their light. Light is a wave form, and its frequency is a bit like the beats of a clock. If you count the beats of the star's light and compare them to the standard beats from light in the laboratory you'll find that the star's light beats more slowly. It's the same with our sun.

If time stands still in a black hole, surely it ceases to exist?

It becomes a meaningless thing. As far as we are concerned, the situation on the surface of a black hole is static.

Yet the inference of light not being able to get out is that something is happening in there but we just can't see it.

There is what we call an 'event horizon' on the black hole which separates events which can be seen, those in the outside universe, from those which can never be seen, those within the black hole.

It's a mistake to think that if you went to the surface of a black hole that somehow you would reach the edge of the universe, the edge of space and time. If you were to go on inside the black hole you would be in a region of space and time that would look perfectly normal to you.

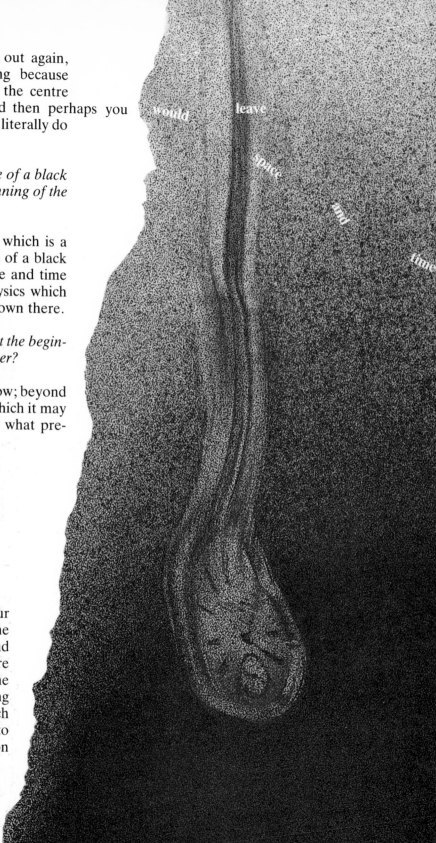

The problem is that you would never get out again, yet you wouldn't be there for very long because you'd be falling rather rapidly. You'd hit the centre of this object after a very short time. And then perhaps you ~~would leave~~ ~~space~~ ~~and~~ ~~time~~

At the centre of a black hole, space and time literally do have a boundary.

Does this edge to space and time at the centre of a black hole have anything in common with the beginning of the universe?

Yes. It's called a 'singularity' in the trade, which is a euphemism for 'we don't know'. The centre of a black hole is a region of the universe where space and time cease to be meaningful. All the laws of physics which make use of the space-time concept break down there.

And an atheist asking a similar question about the beginning of the universe would get the same answer?

Yes. A singularity beyond which we don't know; beyond which space and time do not exist; beyond which it may not be meaningful to ask such questions as what preceded or caused the singularity.

Concepts of space and time are very deeply rooted in our language and it's very hard to talk about anything in the physical universe without assuming the background framework of space and time. Now, those of us who are interested in ideas in areas beyond the beginning and the end of time, want to know if we can say anything meaningful about the sort of substructure from which space and time might be built, and we are trying to formulate a scientific language that doesn't depend on the space and time concepts.

Do you think that the 'In the Beginning' of Genesis is meant to stand for the point before which there was no time?

I think the way that it's phrased in the Bible is very ambiguous. It's not at all clear whether 'In the Beginning' means in the beginning of time, or simply that the universe had been existing for all eternity in an uninteresting state and suddenly God decided to make things happen. This led to a lot of controversy among early Christian theologians, because if God was resting for eternity and then at a particular moment decided, 'I'll have a universe', why did he choose that moment? Why not some other moment, and what was he doing before he created the universe, and so on. Some early theologians, like St Augustine, thought that there was no need to think of time in this way. On the contrary, we could think of time itself being created with the universe, that the universe came into being *with* time rather than *in* time. And this was a remarkable anticipation of modern theories of cosmology – today we also have the idea of the universe being created at one instant along with time. So there is no notion of *before*. To ask what happened before the Big Bang is a meaningless question because there was no *before*. All we can say is that the Big Bang represents the past temporal extremity of the universe, the edge of time. And there may be a future one too, an end of time.

So time as such becomes a property of the universe rather than something that has always been there?

Very much so, and this is what modern physics has shown. We used to think that space was a kind of arena in which things happened, and time was the backdrop against which the activity of these things was gauged. We now think of space and time as part of the action. They can change, move, deform, distort. They are part of the play.

Time and space are tied very closely to matter. As matter changes its properties, its gravitational situation, motions, etc., so it drags space and time with it, distorting them in its own way. We now think of space and time

as very much part of the physical universe. They have their own evolution, and maybe even their own birth and death.

We're talking as if this is the *way that things are: that Newton, for example, thought space and time were 'like that', whereas in fact they're more 'like this'. How much are these ideas 'real', and how much are they in the mind?*

All the physicist does is try to model the world, and the language of these models is mathematics. These models make precise predictions, yet they are still models. Now we have some models which are good and some which are not so good, but this doesn't make them right or wrong. We don't ask whether the Second Law is right, whether it applies to everything. The Second Law is a way of modelling how certain things behave and it has a range of validity. Like all laws of physics, it may not be valid under certain circumstances. Newton's model for space, time and motion works very well for things like machinery but it doesn't describe the motion of sub-atomic particles.

The development which led to the notion of physics as just models was quantum physics rather than relativity. Here, you really are up against it, because it's not meaningful to talk about an atom existing as an independent entity in a well-defined state before you have decided what it is about that atom that you want to measure or observe.

To an extent, all physics is about making models, but I don't think this has been appreciated until recently. People did, and still do think that Newtonian mechanics is there to describe reality. But I don't think that physics has ever worked that way.

So we are dealing in concepts of space and time?

Our models are not mere conjecture. They work well. The time dilation effects in relativity, for example, really happen, in the sense that the model predicts what you will measure.

Some people think that the job of the physicist is to discover *the correct laws* that govern the universe: that the laws which we have at the moment work fairly well, but they are not the right ones. I don't see it that way, because I don't think we can ever really answer what is going on 'out there'. Physics is about making models, some of which are better than others.

Yet these models have awesome practical applications?

Yes. Nuclear weapons have arisen from developments in theoretical and experimental physics.

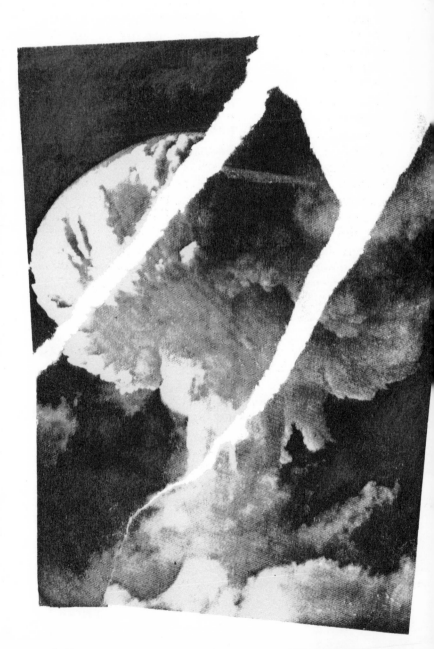

The contributions made by science/technology in World War II – as typified by radar and the Atomic bomb – have resulted in an unjustified status for the wisdom of scientists in the post-war world. The younger generations of scientists have not been slow in picking [...]

The two ev [...] markedly [...] of w [...]

Do you think physicists are responsible for this development?

The problem about being a researcher is that you don't know what you're going to discover until you discover it. If you're for ever scared that you might uncover something that irreversibly changes human fortunes for the worse, then you wouldn't be able to do anything. I think the scientist does have to accept some sort of responsibility, but what we need is a responsible political framework in which people can be scientists, free from the fear that what they might unearth could be exploited by politicians for evil ends.

Yet in these rather fundamental areas of theoretical physics, I don't know many scientists who take the social aspects of what they are investigating seriously enough. Even Einstein was a bit late in condemning the bomb. Having discovered the basic principles in 1905, it was nearly forty years later, when the bomb was made, before he spoke out against it.

The story of J. Robert Oppenheimer, director of Manhattan Project which designed and built the first atomic bombs, is one of the great tragedies of modern science. Oppenheimer, a brilliant physicist and 193[...] radical, sacrificed his frie[...] his ideals, and his person[...] integrity to the demands [...] country in wartime.

**Robert Oppenheime[...]
"conscience-stricke[...]**

He believed it was imperative that America should make the bomb b[...] Nazi Germany could. In 1945, when the first nucle[...] test was held, Germany [...] been defeated and its nu[...] potential was known to h[...] been non-existent. Nevertheless, the new w[...] was tried out on Hiroshim[...] and Nagasaki, targets w[...] Oppenheimer himself he[...] to select. After the war th[...] conscience-stricken phys[...] argued against the build[...] a 'super' or hydrogen bo[...] For this heinous crime h[...] destroyed by the military [...] by some of his professio[...] colleagues. In 1954 his [...] fellow-travelling past wo[...] mercilessly exposed at a [...] trial, and he was depriv[...] his security clearance. Th[...] scientist who was once s[...] have won the war again[...] Japan was now regarde[...] little better than a spy.

A friend of mine has one of those long London gardens with an area of overgrown uncultivated chaos at the bottom of it, beyond the carefully tended lawns and beds nearer the house. One spring evening I found myself out there alone, gazing quietly into a thicket of brambles, bindweed and elder, surrounded by much taller un-pruned sycamores. The sun was setting, illuminating the slow dissolution of vapour trails up in the sky. There was enough light for me to see damp winter browns giving way to rampant greenery, in the midst of which I could make out the bent and rusting frame of a child's swing, choked by the vegetation.

This little jungle had been a play space, and it felt to me then as if all the children who had ever had adventures there were somehow present. The place seemed to ring with their excitement and chatter, their fears, rivalries, secrets and cut knees. All that happened there felt focused in that moment.

Snapping out of this reverie, I started listening to the courtship of blackbirds whose noise had held me subliminally entranced. Looking round I picked out six of them, males and females, perched strategically. Their chorus was loud, drowning the rumble of the receding jets overhead. Their melody seemed to envelop that space in a process of call and counter-call that was beyond communication. The waves of each bird's song seemed enfolded in every other one, creating a synchronous blackbird field. No single bird made sense because each was part of the whole in this collective ritual that would result in more blackbirds.

The accepted wisdom on evolution is a major part of the cluster of ideas that supports the prevailing view of time as linear. Darwinism and neo-Darwinism paint a remorseless historical picture of cause and effect, in which the history of species is strung out, back into the mists of time, along lines of development. We're all familiar with those charts that show how humans evolved from apes, through Homo habilis to erectus to sapiens, carefully dated – four million years ago, two million years ago, 600,000 years ago, and so on.

In his book, A New Science of Life, Rupert Sheldrake throws down a fundamental challenge to the common sense of current thinking on evolution. His hypothesis suggests that biological form is governed by form-shaping fields rather than genetic programming, and that these fields act invisibly across time as well as space. In a strange way, these ideas substantiate the feelings I had about the child's swing and the field of 'blackbirdness'. More significant are the parallels between these ideas in biology, the theory of the Implicate Order that grew out of David Bohm's physics, and the theory of the collective unconscious that Jung developed in his psychology. The fact that different 'scientific' fields are arriving at similar hypotheses suggests there might be something in it.

RUPERT SHELDRAKE was a scholar of Clare College, Cambridge, where he read Natural Sciences. He spent a year at Harvard University studying Philosophy and the History of Science before returning to Cambridge to take a Ph.D. in Biochemistry. As Fellow of Clare College and Director of Studies in Biochemistry and Cell Biology from 1967 to 1973, and as Rosenheim Research Fellow of the Royal Society, he carried out research on the development of plants and the ageing of cells. He is currently Consultant Plant Physiologist at the International Research Institute in Hyderabad, India. He is the author of *A New Science of Life*, first published in 1981 (Blond and Briggs).

193

Where do you stand in relation to Darwin?

Darwin's theory of evolution used to be called 'transformism' because it was based on the idea that one species changed into another by transformation. His main contribution was to point out the role of natural selection in this process. Now I think that this theory is all right as far as it goes, except for the fact that Darwin believed these changes to happen in gradual stages. Today, biologists are disputing this, pointing to the fossil record which reveals sudden jumps. Probably both took place.

But Darwin wasn't sure *how* the evolutionary process worked. Although he had his own theory of acquired characteristics, he didn't know how heredity took place.

> THE PRESENT ORTHODOX THEORY OF HEREDITY, BASED ON DARWIN, IS CALLED NEO-DARWINISM. IT SAYS: ALL HEREDITY IS DUE TO D.N.A., A CHEMICAL CARRIED IN THE GENES; ALL CREATIVITY IN EVOLUTION (EVOLUTIONARY CHANGE) IS DUE TO RANDOM MUTATIONS IN THE GENETIC MATERIAL (D.N.A.); AND, THERE IS NO SUCH THING AS AN ACQUIRED CHARACTERISTIC.

Now, I differ from this conventional view in several respects.

What's the problem? I thought it was generally accepted that the D.N.A. in my cells contained an encoded blueprint of the shape of my hand, foot, or whatever.

That's a very comforting thought, but in fact we understand far less than we usually assume.

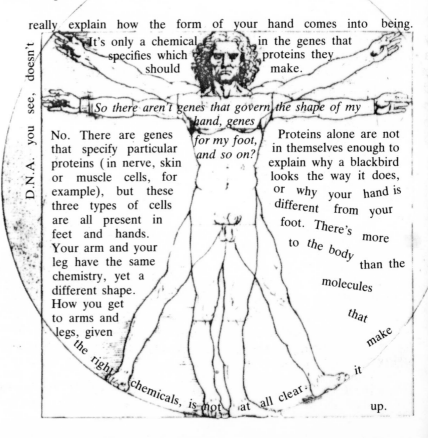

D.N.A. you see, doesn't really explain how the form of your hand comes into being. It's only a chemical in the genes that specifies which proteins they should make.

So there aren't genes that govern the shape of my hand, genes for my foot, and so on?

No. There are genes that specify particular proteins (in nerve, skin or muscle cells, for example), but these three types of cells are all present in feet and hands. Your arm and your leg have the same chemistry, yet a different shape. How you get to arms and legs, given the right chemicals, is not at all clear.

Proteins alone are not in themselves enough to explain why a blackbird looks the way it does, or why your hand is different from your foot. There's more to the body than the molecules that make it up.

What are you proposing?

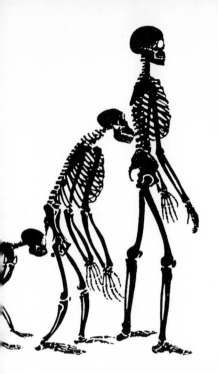

I'm following up a suggestion that's existed for sixty years in biology: the idea that the form of an organism is governed by a form-shaping field, which is called a morpho- genetic field.

It's like an invisible mould.

Now each species has its own field which developing organisms of that species tune into. And within each species there are variations, so in the dog field we'll find sub-fields for greyhounds, poodles, and so on. *What are these fields?*

A field is an invisible spatial structure. No one's ever seen a magnetic field, for example, and you can't touch one. But we see their effects because they can mould physical matter. A morphogenetic field is a field which shapes an organism as it develops.

So D.N.A. does not determine form?

No. It would be the same as saying that the image on your T.V. screen is programmed in the transistors and chips inside the set. Of course, without these components the set wouldn't work, but the pictures don't come from them. In a similar way, D.N.A. is necessary to tell genes to make particular proteins, but the shape of your hand is not programmed inside your genes.

What does determine biological form then?

Most biologists think that with more research we will finally be able to answer this in terms of D.N.A. and proteins. But this is only a hope.

Nobody really knows ●

Where do these fields come from?

I'm suggesting that living organisms pick up on influences coming from outside them, rather like T.V. transmissions. To be tuned into these transmissions, the organism has to have the right genetic material (D.N.A., proteins and so on), just as there's no T.V. picture without transistors and printed circuits.

The difference is that whereas T.V. transmissions are coming from a transmitter that's distant in space, the morphogenetic fields that I'm talking about may be distant in *time* as well as space. They are a transmission from the actual forms of past members of a given species.

So, all dead herons, for example, whether they died ten or a thousand years ago, are busy transmitting a morphogenetic field that heron embryos pick up on today?

Yes. And in the case of living fossils, this influence can span a million years. And living herons are also transmitting these influences.

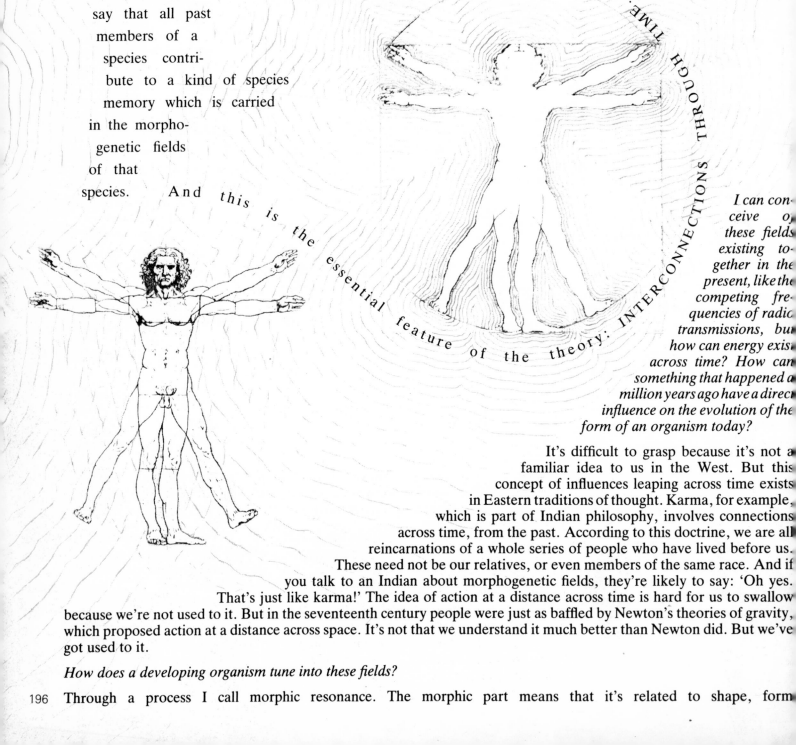

How can a force act across time?

It's not exactly a force.

It's more like a pattern or mould. We could say that all past members of a species contribute to a kind of species memory which is carried in the morpho-genetic fields of that species. And this is the essential feature of the theory: INTERCONNECTIONS THROUGH TIME.

I can conceive of these fields existing together in the present, like the competing frequencies of radio transmissions, but how can energy exist across time? How can something that happened a million years ago have a direct influence on the evolution of the form of an organism today?

It's difficult to grasp because it's not a familiar idea to us in the West. But this concept of influences leaping across time exists in Eastern traditions of thought. Karma, for example, which is part of Indian philosophy, involves connections across time, from the past. According to this doctrine, we are all reincarnations of a whole series of people who have lived before us. These need not be our relatives, or even members of the same race. And if you talk to an Indian about morphogenetic fields, they're likely to say: 'Oh yes. That's just like karma!' The idea of action at a distance across time is hard for us to swallow because we're not used to it. But in the seventeenth century people were just as baffled by Newton's theories of gravity, which proposed action at a distance across space. It's not that we understand it much better than Newton did. But we've got used to it.

How does a developing organism tune into these fields?

Through a process I call morphic resonance. The morphic part means that it's related to shape, form or pattern. And the resonance part means that there's an automatic influence of like upon like. It's a principle of similarity reinforcement.

Like telepathy?

Yes. But across time. It shows most clearly in instincts.

So when I was ten minutes old, rooting for my mother's breast, I was instinctively tuning into the behaviour patterns of past babies. Likewise, the form of my hand tuned into the form of previous hands.

Yes, and it might even have been influenced by what previous hands did. A good example of this is the protective pad on camels' knees. It makes sense to think that through kneeling down a lot camels get these pads where their skin is abraded by sand and rock. But the fascinating thing is that baby camels are born with them.

Now, the conventional neo-Darwinist theory would say that a random mutation came about by chance so that some camels developed pads in just the right places. And because these camels would have been able to kneel down sooner and more effectively, natural selection would have favoured them above camels without pads.

Now, I'm suggesting that through many generations camels have acquired these pads by kneeling down, and the pads became a kind of habit. So as the camel embryo grows in its mother's uterus, even though it has never knelt on sand, it develops the pads through a kind of memory of where the wear and tear came for previous members of the species: through morphic resonance it picks up on the morphogenetic fields of previous camels.

I'm still unclear about how a camel embryo can remember across time.

It depends on the way you view time. The conventional linear models conceive of time as space: so many units back into the past, so many units forward into the future. And even in ordinary speech we speak of the 'near' future, or the 'distant' past. This is perfectly all right as far as it goes, but we tend to lose sight of the fact that it's only a metaphor. Perhaps our primary experience of time comes not so much from theory as from memory, and memory is not connected with time in any obvious linear way. For old people, memories of childhood are often more vivid than memories of the day before and my memories of ten years ago aren't necessarily dimmer than my memories of one year ago. It's as if memory allows us to tune into the past in a way that's not necessarily tied to how long ago an event took place.

So just as I might vividly remember a trip to the sea in my childhood as if it were yesterday, a developing organism remembers in physiological terms how organisms of the same species formed themselves in the past.

Yes, but not as consciously as you suggest. We have two kinds of memory: one is conscious memory where we remember particular events, like the day I fell off my tricycle as a child. The other is unconscious memory, like remembering how to ride a bicycle. When I ride a bicycle I don't remember how I've done it before. I don't have to think about it. All my experience of bicycle riding has blended into the act of doing it,

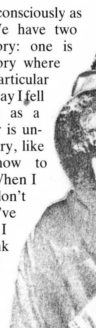

and is unconscious. It has become a habit. Morphic resonance is much more like habit. Like an instinct, it isn't conscious. When a bird builds its nest, it doesn't stop and think, 'How did my ancestors do it?' It just does it. Morphic resonance is much more like the habitual side of our memory, but it involves memories that aren't particular to the individual organism. Let's think about human memory again. We normally assume that memories are somehow stored in the brain. But, in fact, nobody knows this for certain, and the search for the site of memory has been going on

for a long time. Now it's possible that memory may not be stored in the brain at all. Our brains may be more like tuning devices.

So our memories might exist outside our heads? In a collective pool as it were?

We're more likely to pick up on our own individual memories, because we're most similar to ourselves. But it's possible that we could pick up less specific memories from countless people.

Morphic resonance sounds like a biological version of Jung's 'collective unconscious'.

Jung thought that there was a collective human memory that we all tune into. This 'collective unconscious' contains what he called archetypes, basic patterns of experience and ideas, which we inherit from the whole past of our species. Morphic resonance might explain how that inheritance happens.

Now, I'm suggesting that this collective memory is not confined to the psyche, but that it also embraces instinct, and includes the development of biological form, not just in humans, but in all living organisms.

So what Jung discovered in his psychological investigations from dreams and mythology, is one aspect of collective memory through morphic resonance.

Have any experiments been done to confirm the existence of morphogenetic fields?

There is circumstantial evidence from several areas of biology and psychology. Perhaps the best so far comes from a long series of experiments in which rats were put into a maze. Out of the first ones that were tested, each made over 250 errors on average, before they learned how to escape. But as testing went on, subsequent rats improved, until they were learning over ten times faster. But the point is that this improvement was also observed in rats of the same breed that were separated by great geographical distance. These had neither been in contact with the maze-tested rats, nor were descended from them. It appears that they have somehow learned across space and time.

Of course, to test such a radical hypothesis as this, further experiments will be necessary, and some are already under way.

There seems to be quite a similarity between what you and David Bohm are saying.

What we are saying is very compatible. Bohm suggests that the world we see and experience, what he calls the Explicate or outfolded order, is dependent on an invisible underlying structural reality which he calls the Implicate, or enfolded order. And this Implicate Order is not in space and time as we know it, but rather lies behind it, and is the source of space and time as we experience it.

Now, if it were just that, I think what we'd be saying would be rather similar to Plato's philosophy which proposes that there's a timeless order from which this world comes as a kind of reflection. But Bohm is not saying that. He's suggesting that the world we see and experience, the Explicate Order, influences the Implicate Order. There's a kind of feedback from the Explicate to the Implicate Order. So what happens here, like a rat learning a new trick, for example, will feed back into the Implicate Order, so that the next time rats try to learn the same trick somewhere else, they will be influenced by that modified Implicate Order, so learning the new trick more quickly. The Implicate Order, then, provides a kind of connection between now/here and later/somewhere else.

These ideas seem more metaphysical than 'scientific'; more a belief system than something provable.

I think that all scientific ideas have a metaphysical structure to them. Take the 'belief system' of the conventional scientist who claims his theory to be based on proof through hard evidence. What lies beneath that? The basic assumption of the conventional scientific model is also that there's an invisible reality underlying our physical universe. And this invisible reality is largely

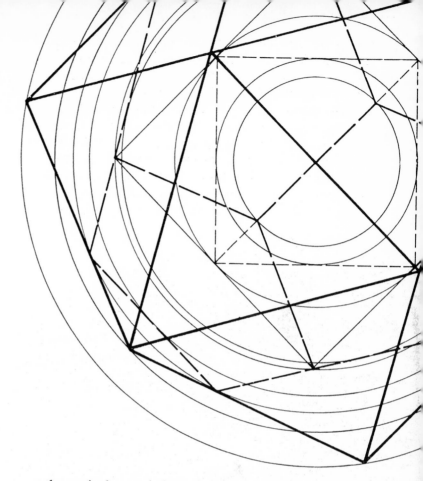

mathematical, consisting of things called the 'laws of nature', which are alleged to be eternal and omnipresent, so that nothing in the universe can escape them. Now, the metaphysical nature of this assumption can be exposed most clearly by asking the hard-nosed mechanical scientist where the 'laws of nature' were before the Big Bang. If they came into existence with the Big Bang, why did they take the form they did? If they were there before the Big Bang, what kind of things are they? They can't be physical, since the physical universe didn't exist before the Big Bang. So they must be metaphysical in some sense. In fact, the 'laws of nature' take in most of the traditional qualities of God. If we look back to the Middle Ages, we find that these ideas grew out of the Christian theological tradition which thought of laws of nature as ideas in the mind of God.

So when David Bohm proposes the Implicate Order as an invisible reality underlying the world, it sounds metaphysical, but when we think about, it's no more so than the conventional scientific view.

Often I'm so driven by the sheer density of ambitions, appointments and deadlines (like the one on this book), that I lose my sense of the present. Locked on to the imperative of tomorrow, I become desensitised to my immediate environment so that even at the most private and potentially contemplative times I find it impossible to do nothing. Sitting here on the lavatory, for example, I'm desperate for reading matter so as not to face myself, which means I'll have to reread that old colour supplement I already know virtually by heart.

In addition to the complex madness of its schedules, with its sharp demarcation of work time and leisure time, contemporary industrial society conspires to deny us the present in all sorts of ways. At its heart lies the sacred dictum of 'economic growth', which constantly shifts our attention from the reality of here and now towards the hope of a better future. As Mike Cooley has expressed earlier, the pace of technological change that sustains economic growth creates a climate of constant de-skilling and redundancies in which one is either consigned to the negative timelessness of the dole queue, or the desperation of trying to keep up.

The ideology of consumerism is central to this process: more, bigger, quicker and faster are the watchwords that soothe the alienated present with the usually disappointed offer of things. Our experience of time becomes a line teased out from today's anxiety towards the unfulfilment of tomorrow.

Flicking through the pages of the colour supplement on my knee, I can see how advertising fuels this neurosis with its shrill cries to buy now before it's too late, its final offers and latest trends, dressed up in the one-step-ahead language of ultra modernity.

In denying the present, consumer society denies life. Distracted from the moment, we are unable to expand into it: the stillness of contemplation, the timelessness of loving, the infinite depths of ecstasy . . . are squeezed by the encroaching brackets of future and past. As if aware of this argument, advertising is now colonising this space.

Benson and Hedges advertisements, for example, abound with symbols of eternity and timelessness. Hour glasses, fossils, pyramids, gold ingots and sand combine a deliberate imagery of time that plays on our deepest spiritual needs. These needs gain the illusion of being gratified through the smoke-screening of future and past. Nicotine numbs difficult emotions. Less successfully, we are offered enlightenment by the coupling of smoking with the suggestion of the third eye when it's 'time to enjoy Silk Cut'. John Player, also, co-opt the language of time with weak punning in phrases like 'Black in Time'.

Drink advertising invades the same terrain, with its use of predominant blue – the colour of eternity – and tongue-in-cheek exhortations to 'free the spirit'. While car advertising has appropriated our sense of the life process itself. As if to give natural legitimacy to the ecologically disastrous concept of planned obsolescence, Volvo now has 'a son', the Renault 25 is 'the origin of a species', and Volkswagen is 'born again' and 'quick to reach sixty, slow to grow old.'

The expectancy created by the cumulative process of news also draws us out of the present into the future. Waves of events top each other as news 'breaks' in the morning papers, the evening papers, and at the hourly markers of news bulletins. Bombarded with the heightened melodramas of 'latest developments', this overblown sense of history nags us forward in a public linear rush, virtually stigmatising the intimate personal present of just being.

As if to counterbalance this frantic pull into the future, the past now tugs in the opposite direction. Nostalgia seems to grow in direct proportion to anxiety, as a gentle placebo. These emotions are extremely vulnerable to good marketing, as indicated by the incessant production of the packaged past: in fashion revival, music, cinema, costume drama and even morality. Stretched between future and past, the present becomes a fidgety vacuum of premature strokes and heart conditions.

I am still sitting on the lavatory desperately distracting myself from my fear of vacancy. I've finished the colour supplement, scanned an old business news, and even craned down to read the label on my trousers which are down round my ankles. Now, I'm ashamed to say, I've descended to reading a pound note to sustain the activity.

I'm already bored with the Queen and the Chief Cashier's signature, but who's this strange type on the back, with his telescope, prism and Principia? Why, it's the much maligned Sir Isaac (born 1642, died 1727), next to some visualised Law of Motion, which if we were to see it written down algebraically would be sure to embody the 't' of Newton's absolute linear time. Could this be the final proof that time is money? Chatter . . . chatter . . . chatter . . .

JILL PURCE is the author of *The Mystic Spiral: Journey of the Soul*, and is General Editor of the 'Art and Imagination' series for Thames and Hudson. She received a Research Fellowship in the King's College Biophysics Department after completing her degree in Fine Art. Her interest in the spiritual and magical properties of music and the voice has led her to Germany to study with Karlheinz Stockhausen, and further afield to learn Mongolian and Tibetan overtone chanting with the chant-master of the Gyöto Tantric College. She gives lectures and voice workshops throughout the world.

JILL PURCE –
ANXIETY, TIME
AND SOUND MEDITATION

What do you mean by 'the chattering mind'?

It's what the Tibetans call '*namtok*', and it arises from the sense we have in everyday life of me 'in here' and the world 'out there'. This division between inside and outside is the basis of language and the use of the rational mind. For example, if I look at a tree, I might automatically find myself thinking: 'That's a tree, and it

reminds me of this, which reminds me of that, and tomorrow I must . . . ' and so on. This is the chattering mind, which we're at the mercy of unless we can find some way of interrupting the cycle of perception and conceptualisation. It prevents us from being 'present'.

I use sound meditation, based on Tibetan chanting. Now, sound, more than anything else, might be felt to be coming from 'out there' with us 'in here' listening to it. But it has this quality that enables us to integrate outside with inside so there's no longer any division. It's possible to hear sound in such a way that you integrate with it and interrupt the chattering.

In the workshops that I give we try to find our own sound, the sound which we feel most comfortable with, which is a fairly low sound. And when we've found it, we try to enhance it and bring out its harmonics as separate notes present simultaneously with it.

What is a harmonic?

Sunlight is made up of the colours of the rainbow. You can see this when you take a prism and reveal the spectrum. The harmonics of a note are similar. If you play a note on a horn it will contain a number of other notes. These are the harmonics of the horn sound which give it its particular colour, a colour which distinguishes it from an oboe, or your voice. The different shapes of instruments amplify different harmonics. This is why the same note played on different instruments has a characteristic colour or quality – it's called *timbre* in music.

The human voice is similar: the difference between the vowel sounds results from the emphasising of different harmonics. Our language is based on these distinctions, permutated with the percussive, noise sounds of the consonants which punctuate them in various ways.

Now, it's possible to emphasise these vowel sounds in such a way that you don't just hear the basic note, but also the harmonics that colour it. By modulating vowel sounds in a very controlled way, it's possible to reveal the spectrum of sound within sound. So above the fundamental note you hear extraordinary flute-like sounds.

How do these techniques stop the chattering?

Using the voice in this way, the aim is to integrate with the sound. By listening to the sound right through, from its absence, keeping your attention on the transition from absence of sound into sound, actively listening throughout the duration of the sound, and finally paying attention to its decay, back into the absence of sound once more . . . In this way you are present for the whole duration. You must listen to the sound as you make it, so that you're in a state of continual awareness.

So you are trying to create an eternal presence through sound?

Yes. And for this reason you get a sense of timelessness. Time becomes remote, and the sensation of its passing no longer exists. You have the sense that there are no distinctions between you and the sound, nor is there any sense of duration.

Chanting, or making the sound yourself, is not the only form of sound meditation. In one very ancient technique, the instruction is: 'Listen to the sounds around without identifying them.' For example, I might hear a car outside. Now, my normal response would be to name it in my mind, and this would evoke all sorts of associations, which would set my mind off on its normal pattern of chattering. Instead, I listen to that car sound with increased intensity, but without identifying it as 'car'. So I don't just hear it, I use it. I don't think of it as an intrusion and go somewhere where I can't hear it. I simply don't name it. I integrate with the sound itself by disconnecting it from the concept of its source. But then, as soon as I call it a car, I start to think: 'Well, I've got to go out soon . . .' and I'm no longer in the present.

Because we know the sound of a car, the familiarity makes it hard to break the cycle of perception and conceptualisation. Less familiar sounds, like Japanese temple gongs, aren't so easy to name.

Another way of using sound to interrupt the internal dialogue is by a very sharp and sudden sound of the voice. This ancient technique leaves you in a state of 'illuminated astonishment', and if you can remain in that state with a sense of presence, you can begin to experience the nature of the mind itself beyond thought. You interrupt the thoughts which are linked to time and open the way out into timelessness. There's a direct relationship between our sense of time and the activity of our mind in thought.

In a sense, interruption is the basis of all music: sound and the absence of sound. It's this initial polarity which gives rise to what we might call the continuity of musical timing: the idea that all music derives from the interruption of the continuity of our experience. The slowest interruptions give us first duration, then rhythm.

Rhythm is the simplest way in which music affects us because our own bodies have a multiplicity of rhythms for external rhythms to resonate with or entrain, from the rhythms of our breath, heart and brain down to those of every cell, molecule and atom. The loudest of our rhythms is the heartbeat: if you listen to the heartbeat you get lulled into an almost primal state. Rhythm is the most obvious way in which sound affects the human being. Then, if you speed rhythm up beyond sixteen c.p.s. you start to hear the interruptions as continuous tone.

Melody, and the use of tone, also affects us emotionally. The ancient Greeks called this *ethos*, where different keys produced a different mood – sadness, happiness, anger, and so on.

And then we come to the finest vibrations, the harmonics within the tones, where the effects are at their most subtle. This is one of the reasons why I use harmonic chanting: the more subtle the vibrations, the deeper and more powerful the effect, until we can work on what is called in the Hindu tradition 'the subtle body', the 'energy body' which some traditions think is around the physical body. The harmonic chanting becomes a kind of homeopathy of sound.

A kind of self-healing?

Yes. This is really the use of 'mantra'. In the ancient Hindu tradition, Shiva created the fifty-one letters of the Sanskrit alphabet by his drumming. The letters actually fell out of his drum. Now the 'mantric sounds' in this tradition are all the sounds of the Sanskrit alphabet, and each was allotted its particular place on the body. And by making a particular sound, you opened up that part of the body.

The original sound was the sound of the 'OM'. It is the sound of the kinetic stress produced by the original movement in the cosmic stuff. If you listen to the 'OM' when it is chanted correctly, you can hear all the harmonics emerging from it. It is made of three sounds: the 'A' sound, the 'U' sound, and the 'M' sound. A . . . U . . . M: OM. And these sounds correspond to the Trimurti, the three Principals of the Universe – Brahma, who created the world; Vishnu, who preserves it; and Shiva, who with the letters that fell from his drum, created language, and therefore thought and the rational mind.

As the letters fall out of Shiva's drum, they land within each of us. This means that each of us has all the creative sounds within. So the real use of 'mantra' – like chanting the 'OM' in meditation – is like an echo. A human re-creation of the original creative sounds. It is the nearest we can get to the original sound of 'the thing itself'. By unfolding these sounds within the body we are making an internal spiritual journey, 'decomposing' the hardened habitual patterns of our lower nature.

Many traditions carry the idea of the creation of the world through sound. In the Gospel of St John we read: 'In the Beginning was the Word, and the Word was with God, and the Word was God.' Now, we find exactly the same description in the Hindu Veda: 'In the Beginning was the Nada, and the Nada was Brahma . . .' and so on.

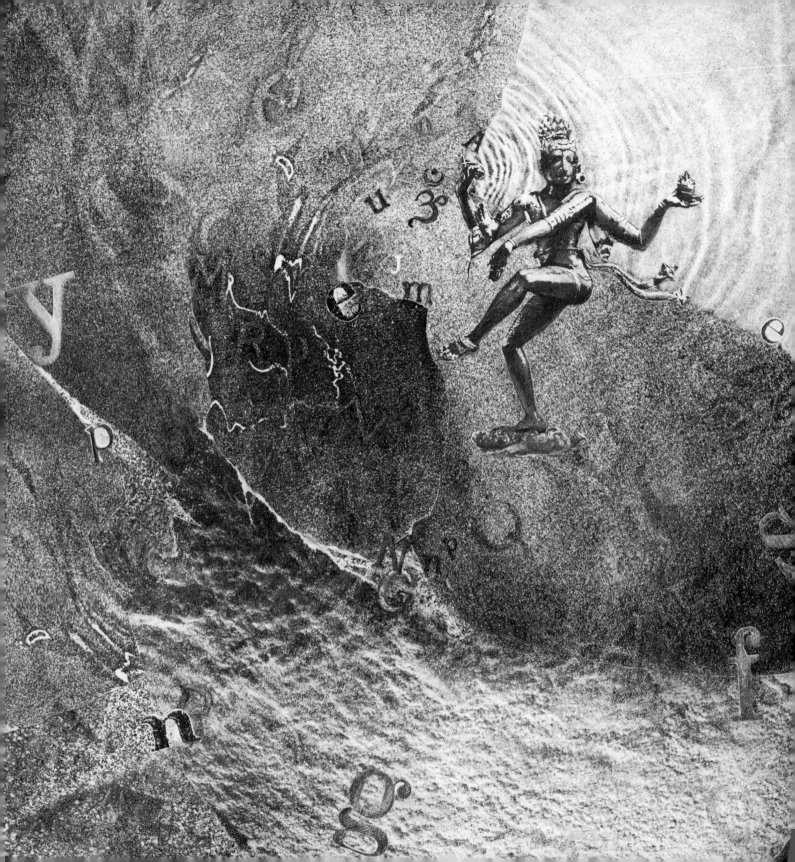

In the Western tradition we have a very different view of the spiritual journey. Instead of it being an internal journey through the body, as in the Yogic tradition, it's an external journey, across space and time. But sound is still central here. The neo-Platonic tradition inherited by the Church fathers describes how, when human beings are born, their souls have to traverse the vast and immeasurable distance that separates God from the Earth. The soul makes a journey downwards from God through all the hierarchies of being – from Seraphim and Cherubim to Archangels and Angels – passing through the levels of mind and so on, one by one through the planetary spheres, down through the elements, and finally to Earth where it is born, incarnate in the body. Now, as the soul passes through each planetary sphere, it picks up the quality of the planet in proportions according to the particular moment in space and time. Each planetary sphere also has its own note, and the soul picks these up too. This is 'the music of the spheres'. When the soul is born it is constituted of the sounds of the planets as much as their other qualities. When we die, our soul makes the reverse journey back to the planetary spheres, and on its way it relinquishes qualities and sounds that created it.

I think most people would find this an effective metaphor, but they wouldn't accept it's how things really are.

There's a very old story that goes back to about 100 B.C. It's from Apollonius and is about Philolaos, a pupil of Pythagoras. He describes how one day a shepherd, returning home to his village, heard singing coming from the tomb of Philolaos. Terrified, he ran back to the village and sought out a disciple of the philosopher called Eurytos, and told him: 'I just heard a note coming from the tomb.' Delighted, Eurytos asked: 'In what key was he singing?' You see, he wanted to know where on the return journey his teacher had got to.

Nice story, but nevertheless a myth?

Can we ever know how things *really* are? In the Eastern tradition these things are still very real and go beyond metaphor. In the Yogic tradition, the sounds actually exist within the body. Working with the mantric sounds affects the body quite precisely.

Why do you think that the traditionally Eastern practices of Yoga and various forms of meditation have attracted so many people in the West in recent decades?

In the East they have a more developed and systematic understanding of the processes of the mind, which has evolved through many generations of Yogic and other traditions. This doesn't exist in the West where psychology is barely a century old and religion is seen far more metaphorically than scientifically. The East fills an enormous gap in this respect.

Why hasn't the West developed this?

Different cultures develop different qualities. The Western tradition has come to emphasise the external world which it seeks to understand in a linear way. Whereas in the East, the development was more an internal one, seeing the process within the body itself.

Do you think the tradition of Western classical music reflects the linear way we seek to understand the world?

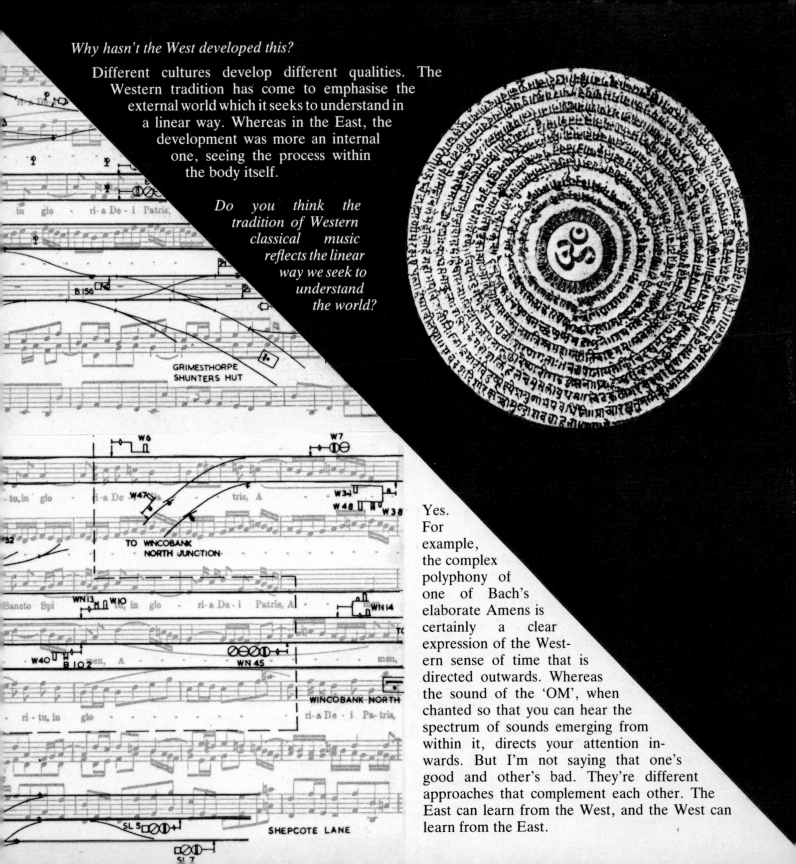

Yes. For example, the complex polyphony of one of Bach's elaborate Amens is certainly a clear expression of the Western sense of time that is directed outwards. Whereas the sound of the 'OM', when chanted so that you can hear the spectrum of sounds emerging from within it, directs your attention inwards. But I'm not saying that one's good and other's bad. They're different approaches that complement each other. The East can learn from the West, and the West can learn from the East.

CLOCK OF CLOCKS

I

Toothed wheels, the dust ringing
With tock, belfry of clappers
With high long tones, like tears,
With skirts that cry intolerably
And wheels packed between four corners
Of stone whose inner crystals
Take o'clock's shape. I tie it
To my wrist, I pocket it, the case
Ringing like the tiles, the rafters,
A church on my wrist, a box of hours.
Who do I meet in this crowded
Church, with the pews meshing
And the greasy altar with a wheel hanging
And the rose-windows that connect
And the doorway like a toothed almond?
Is it a gymnasium, with forfeit mincers?
It is a tree of steel I am in, or a smithy
Full of anvils arranged on wheels
That beat with hammer-blows, or a stable
Full of chariot-wheels that cannot stop fighting,
And clockwork hooves galloping,
Of empty goods-trains running on time?

II

The clock of oil, through which
The shocks run in shadows;
The clock of wood, from which the fruit
Drop every second; the Chinese wooden
Water-clock which fills a mill
Swarming with woodlice like cunningly jointed
Bracelet wrist-watches with legs;
The sun-house clock
Of revealed architecture through whose spaces
Shadows and lighted areas fall, creating
Rooms, cabinets and clocks, staircases,
And deleting them as the day passes, containing
The cobweb-clock that has eight hands
And jaws and feeds on motes; the two-door clock
That sends rains in the hands of a young girl
Holding cat o' nine-tails of silver chain,
Or sunny weather like a man

With a painted torch-bulb for a face;
The unfrocked priest says mass
In the roofless clock, under the diamond points
Of the clock of lights whose teeth are suns;
On the marble mantelpiece the flint clock
Strikes sparks every second of the day and night,
Invisible by sunlight, whose pillared halls
Crackle with earth-fire, the true time.

III

The clock of meals, toothed wheels
With soft lips, eat
And time stops; the clock of sex,
That galloping timepiece; the clock
In the burning house that ticks on
Until it melts; the clock of cogged soot
In the hooded smoke of bonfires burning seconds
That drop from rusty clocks striking autumn;
The blood-clock of a scarlet ape that bleeds
Exactly at new moon on the temple straw;
The white snow-clock-hat that fits the mountains;
The joke that is a well-timed clock; the clock
That is an oak-tree dripping with water-seconds,
Bright ticks of water; and
The clocks made of money, the clock-exchange;
The clock of shoals of fish under the moon
And wooden fishing-boats with the trawls out
At the right tide to pull up the silver
Loads of time under the moon, which is a blank clock-face
That has sent its hands to sweep the tides; the hand-written
Clock-face of a summons; the clock
Of aid-to-the-poor; the Arts Council
Like a clock with pockets and in them wallets
And when it chimes, notes slide out of the pockets;
The clock of grannie's lace that dishevels
In the tomb that undoes all her knots;
The clock of a man counting his billion seconds
Each one of which is as red as corpuscles;

And the clock of clocks itself, once
Passed through like a hoop, once back.

Peter Redgrove

How would you describe the time of poetry?

A continuum. Poetry is trying to create a trance in which the attentive reader or listener sees visions. A good poem is a highly formed dream. You're trying to induce dreams in people. After a poetry reading nearly everybody dreams vividly. That's to say, one has the poet's dreams, retaining in the memory vivid images and sequences which belong to the poet.

Do you find that poetry enables you to express certain ideas about time?

It allows you to play with time, just as cinema plays with time. I think poetry is very much like film in that way. In the poem 'Clock of Clocks' I was thinking how very regulated our lives are by measure, including the menstrual measure of course, and that most things can be seen as clocks that control us all. I simply wanted to be funny about this.

A poem is very much like the network of pearls or the spider's web full of beads of dew, where each jewel or bead of water reflects every other one. It's the holographic analogy again. Writing a poem is very much about every part reflecting every other part.

This is what poetry is trying to do of course, to make a whole thing which is different from ordinary speech. A poem is a kind of festival of speech.

In my dream, so to speak, I experience the whole of a poem simultaneously. The difficulty is that in writing it, you are trying to bring it out of that simultaneity into time.

The notion of an order beyond, or an underlying reality, has been a recurrent theme of the last part of this book, and it sounds very much like a euphemism for God – or the goddess, gods and goddesses, the Atman, Brahman, the Tao, the Absolute Principle, or whatever other names are given by different religions to the ultimate mystery. This is how different contributors have expressed it:

Peter Redgrove: 'The old legends in Cornwall say that the gods need us as much as we need the gods.'

Danah Zohar: 'I think of us as God's partners in evolution. He needs us on the world stage. Our petty temporal acts are in fact weaving this enormous web. This is what morality is about: we emerge from the Implicate Order (or the well of being or vast sea of potential, whatever you want to call it), and I think we have the moral responsibility to make something out of life, while we live it in an individual state, such that whatever it is we return, makes the fabric of that well of being just that little bit richer.'

Rupert Sheldrake: 'Bohm is suggesting that the world we see and experience, the Explicate Order, influences the Implicate Order. There's a kind of feedback . . .'

In whatever way it's put, the implication here is that human psychic evolution is an indivisible aspect of an evolving god. Far from being able to defer to an entity beyond the edge of time, we find ourselves as responsible to it, as it is to us. In fact we are inseparable from it: we are its manifestation, and it is our collective spiritual essence.

These concepts have been the shared reality of mystical experience across many cultures for thousands of years and are referred to as the 'perennial philosophy'. In one form or another they lie at the core of most religions, yet won't be tied down to any orthodoxy. In the light of the 'New Physics', this might be termed a quantum view of religion, in which it is no longer possible to separate god-humanity the observer from humanity-god the observed. Far from offering the comfort of a spiritual insurance policy, this faces us with awesome responsibility.

It has been suggested earlier that in the Implicate Order – the underlying reality of the holographic universe – all time is simultaneously present. Paradoxically this view does not contradict our experience of life and the universe in the Explicate Order, which appears to have a direction in time. Much of this book has been concerned with substituting a complex view of time for the dictatorship of linear time. In place of one view of time, which moves in a straight line at the same rate everywhere in the universe, we've been posing an infinitely faceted time which circles, falters, zig-zags, appears to stand still, spirals, sidles and rushes at different levels of process and experience. Yet the overall direction this complexity seems to have is forwards.

Nor is the idea that all time is simultaneously present the deterministic world-view it appears to be. It does not mean that a blueprint of the future somehow exists in the present, precisely conditioned by the past and invulnerable to free-will. The future is present, but as a sea of potential which human beings – among other entities? – will collapse into certainty through their actions. Freedom of choice, although often severely limited by the constraints of our differing histories, is none the less real. Acting on that future is, as Liz Greene puts it, like 'learning to dance' – a reconciliation of the moral choices posed by our freedom with the limits of what is possible.

The consequences of human free-will are not necessarily limited to affecting our own futures, or future events in this small region of the galaxy. The suggestion here is that it shapes, and is part of – perhaps along with other life forms in the universe – the evolution of a greater 'something' – let's dispose of the loaded word 'god' here – in which we are implicated, yet which transcends us in the evolving cosmos.

This 'bootstrap' theory of religion confronts us with a terrifying existential reality. We are no longer the creation of a supreme being, or the poor mirror of some original abstract perfection. We're more like the fragment of a hologram, tumbling forwards with that whole something within us, helping it to define itself, as we define ourselves, pulling ourselves up by our bootstraps.

Christopher Rawlence

212

SOURCES AND SUGGESTED READING

Once Upon a Time

BERGER, JOHN. *And Our Faces, My Heart, Brief as Photos*, Writers and Readers, London, 1984.

BERGER, JOHN. *Another Way of Telling*, Writers and Readers, London, 1982.

CALVINO, ITALO. *Italian Folk Tales*, Penguin Books, London, 1982.

Time is Money

ATTALI, JACQUES. *Histoire du Temp*, Fayard, Paris, 1982.

CAPRA, FRITJOF. *The Turning Point*, Wildwood House, London, 1982.

COOLEY, MIKE. *Architect or Bee?: the Human/Technology Relationship*, Langley Technical Services, Slough, 1980.

HOWSE, DEREK. *Greenwich Time*, Oxford University Press, Oxford, 1980.

LANDES, DAVID. *Revolution in Time*, Belknap Press, Harvard, 1983.

THOMPSON, E.P. *The Making of the English Working Class*, Penguin Books, London, 1968.

THOMPSON, E.P. 'Time, Work Discipline and Industrial Capitalism', *Past and Present*, vol. 38, 1967.

ZERUBAVEL, EVIATAR. *Hidden Rhythms*, University of Chicago, Chicago, 1981.

Holy Days

BUSHAWAY, BOB. *By Rite*, Junction Books, London, 1982.

ELIADE, MIRCEA. *The Myth of the Eternal Return*, Routledge, London, 1974.

GREEN, MARIAN. *A Harvest of Festivals*, Longman, London, 1980.

RAWE, DONALD R. *Padstow's Obby Oss and May Day Festivities*, Lodenek Press, Padstow, 1982.

REDGROVE, PETER. *The Sleep of the Great Hypnotist*, Routledge, London, 1979.

Moonshine

CAMPBELL, JOSEPH (ed.). *The Portable Jung*, Penguin Books, London, 1980.

VON FRANZ, MARIE-LOUISE. *The Feminine in Fairy Tales*, Spring Publications, Dallas, 1972.

VON FRANZ, MARIE-LOUISE. *On Divination and Synchronicity: the Psychology of Meaningful Chance*, Inner City Books, Toronto, 1980.

VON FRANZ, MARIE-LOUISE. *Time, Rhythm and Repose*, Thames and Hudson, London, 1978.

GREENE, LIZ. *The Astrology of Fate*, George Allen and Unwin, London, 1984.

HALL, NOR. *The Moon and the Virgin*, The Women's Press, London, 1980.

HARDING, ESTHER. *Woman's Mysteries*, Rider, London, 1982.

MANN, A.T. *Life-Time Astrology*, George Allen and Unwin, London, 1984.

REDGROVE, PETER and SHUTTLE, PENELOPE. *The Wise Wound*, Penguin Books, London, 1980.

WEIDEGGER, PAULA. *Female Cycles*, The Women's Press, London, 1977.

WILHELM, RICHARD (with an introduction by C.G. Jung). *The I Ching or Book of Changes*, Routledge, London, 1951.

Uncertain Times

BOHM, DAVID. *Causality and Chance in Modern Physics*, Routledge, London, 1957; 1984.

BOHM, DAVID. *Wholeness and the Implicate Order*, Routledge, London, 1982.

CAPRA, FRITJOF. *The Tao of Physics*, Wildwood House, London, 1975; Fontana, 1976.

DAVIES, PAUL. *God and the New Physics*, Dent, London, 1984.

DAVIES, PAUL. *Space and Time in the Modern Universe*, Cambridge University Press, Cambridge, 1977.

DOSSEY, LARRY. *Space, Time and Medicine*, Shambhala, Boulder and London, 1982.

EDDINGTON, A.S. *The Nature of the Physical World*, Cambridge University Press, Cambridge, 1928.

FERGUSON, MARILYN. *The Aquarian Conspiracy*, Paladin, London, 1982.

GROF, STANISLAW. *Realms of the Human Unconscious*, Souvenir Press, London, 1979.

REDGROVE, PETER. *The Wedding at Nether Powers*, Ron Hedge, London, 1979.

SHELDRAKE, RUPERT. *A New Science of Life*, Blond and Briggs, London, 1981.

WILBER, KEN (ed.). *The Holographic Paradigm*, Shambhala, Boulder and London, 1982.

WILBER, KEN (ed.). *Quantum Questions*, Shambhala, Boulder and London, 1984.

WILBER, KEN. *Up From Eden*, Doubleday/Anchor, New York, 1981.

ACKNOWLEDGMENTS

The Editor and Publishers wish to thank the following for permission to reproduce visual material:

Academia, Venice, 7; 194; 196; 198
Acropolis Museum, 130 (second left, bottom)
Alinari, 102
A.U.E.W. Foundry Section, 45
Barnaby's Picture Library, 14; 40
B.B.C. Hulton Picture Library, 44; 59 (top right); 69; 73 (top and middle left); 76/77 (top); 80; 81 (top right); 94 (bottom left); 117; 181
B. S. Beckett, 15 (from *The Illustrated Biology* by B. S. Beckett, published by Oxford University Press, © 1978 by B. S. Beckett)
Bibliothèque Nationale, Paris, 36 (bottom left)
Biff Products, 5; 31; 74; 132; 145; 146
S. C. Bisserôt, 24
The Boeing Company, 57 (bottom right)
Jacqueline Botaglissio, 178 (Kurdish woman, bottom right)
The British Library, 101; 127 (bottom left)
——Oriental Department, 7
——Newspaper Library, 159 (left)
The Trustees of the British Museum, 158; 208 (bottom)
——Museum of Mankind, 138
——Natural History, 194/195
Brunel University, Uxbridge, 38 (top)
Basil Buckland, 130 (McGill cartoon)
Cairo Museum, 140 (top left)
Lucien Clergue, 22
Terry Connolly, 89; 90 (effigy, left); 90/91 (top)
Control Data Corporation, 37 (bottom)
Daily Mail, 46; 148 (bottom right)
Daily Mirror, 158
Deutsches Museum, Munich, 40 (bottom left, Vaucanson's duck, 1738)
Express Newspapers Ltd, 70; 94/95; 114; 158; 192
John Freeman, 119
Ford Motor Company, 41
Futures Forecast P.L.C., 127 (top left)
Giraudon, 140 (standing figure of Khons, Moon god and relief of Akhenaton)
Lyn Glassborow, 52
Richard Glassborow, 35; 96
Guardian, 192
Hale Observatories, 18
Mrs Haywood, 34
Gaylord O. Herron, 100 (bottom); 197 (left)
Judith Higginbottom, 110
Illustrated London News, 42 (top left); 43; 50 (left); 53 (top, middle); 56
The Iveagh Bequest, Kenwood House (G.L.C.), 25 (bottom)
The Warden and Fellows of Keble College, Oxford, 48 (W. H. Hunt, 'The Light of the World')
Peter Kennard, 160 (bottom left) ('Haywain with Cruise Missiles', photomontage after John Constable)
André Kertész, 19; 21
Keystone Press Agency Ltd, 179
Kimberley Clark Ltd, 150 (top right)
Alex King, 84 (bottom left, first and second); 90 (bottom middle); 91 (bottom middle)
Langley Technical Services, 39 (bottom right)
Christian Lignon, 139
Celia Lowenstein, 55 (photo, bottom right)

Mansell Collection, 20
Maurits van Nassau Mauritshuis, The Hague, 25 (top)
Mitsu Art Gallery, Tokyo ('The Universe' by Sengai), 135
Jean Mohr, 9; 11
R. A. Moulding Esq. (Oak-apple Day, Great Wishford, 1906), 67 (bottom left)
Maggie Murray, Format, 105
The Trustees of the National Gallery, London, 25 (third from top)
National Galleries of Scotland, Edinburgh, 17
The Trustees of the National Maritime Museum, London (photo: G. Wilhide), 54 (right)
National Museum of Labour History, 70 (top left)
The Hydrographer of the Navy, Taunton, Somerset, 115
Nelson-Atkins Museum of Art, Kansas City, Missouri (Nelson Fund), 169 (centre of montage, 'Tree of Life and Knowledge', Indian bronze)
Lennart Nilsson, 174 (right); 175 (top right) (photographs of foetuses in the womb, published in Great Britain in *How You Began* by Kestrel Books, Penguin Books Ltd)
Palitoy, Leicester, 33 (model of steam engine); 37 (centre); 50 (top and bottom band); 51 (gears); 53 (ships); 55 (bottom right); 59 (second from top); 105 (top left); 106 (bottom left)
Pictor International, 131 (centre)
Planetarium, Northern Ireland, 148
Brenda Prince, Format, 107
Rigby Publishers, Adelaide, 151 (figure, top) (from *The Aborigines and their Country* by Charles P. Mountford, 1975)
Ann Ronan Picture Library, 36 (top right); 208 (top)
Doc Rowe Collection, 66
Royal Astronomical Society, 116 (top right); 123
Sanity, 192 (top left and right)
Lynne Saville, 63 (portrait of Peter Redgrove)
Science Museum, London, 51 (photograph of the portrait of John Harrison)
Science Photo Library, 1 (top)
David Shinn, 68
Posy Simmonds, 60; 71; 72; 99
St Bavo, Ghent, Belgium, 104
Staatliche Museen, West Berlin, 25 (second from top)
Tambrands Ltd, 113 (top left and right)
The Trustees of the Tate Gallery, 172 (bottom left)
Thomas Cook, 73 (top right); 75; 76/77
The Times, 70; 96 (top right); 148 (bottom left)
Trades Union Congress Library, 2, (bottom left); 46 (watch in montage); 55 (bottom left); 73 (watch in montage, top)
University of Birmingham, Physics Department, 92/93; 168
C. Verlucca, 118
The Board of Trustees of the Victoria and Albert Museum, 103; 130 (top left); 133
The Wellcome Institute Library, London, 118 (middle left)
Glenn Wilhide, 33 (the Salisbury Clock and engine wheels); 36 (top right); 37 (top)
Working Woman Magazine, 142 (top middle and bottom right)
Yorkshire Miner, 158 (bottom left)

The Adonis poem on p. 16 was translated from the French by John Berger.

The Production company have tried in all cases to trace the copyright holders of all the photographs used in this book. If there are cases where we have failed to make contact in time for the publication of credits, we apologise and will gladly rectify the matter in future editions.